MW00996536

Picturing Empire

Picturing Empire

Photography and the Visualization of the British Empire

James R. Ryan

THE UNIVERSITY OF CHICAGO PRESS

For Christopher Mark Millington Ryan, 1941–74
and Mark William Ryan, 1997–

James R. Ryan is a lecturer in the school of Geography at
the University of Oxford, England

This book is published outside the USA
in the 'Picturing History' series,
edited by Peter Burke, Sander L. Gilman,
Roy Porter and Bob Scribner

The University of Chicago Press, Chicago 60637
Reaktion Books, Ltd, London W1P 1DE

Printed in Great Britain
06 05 04 03 02 01 00 99 98 97 123456

ISBN 0-226-73233-9 (cloth)
Library of Congress Cataloging-in-Publication Data

Ryan, James (James R.)
Picturing empire : photography and the visualization of the
British Empire / James Ryan.
p. cm.
Includes bibliographical references) and index.
ISBN 0-226-73233-9 (alk. paper)
1. Great Britain—Colonies—Historiography. 2. Great Britain—
Colonies—History—Pictorial works. 3. Photography—Great Britain—
Colonies—History 4. Imperialism—Historiography. I. Title.
DA16.R93 1997
941.08—dc21 97-26401
CIP

This book is printed on acid-free paper.

Contents

Acknowledgements

Many people have been generous with their time and knowledge during the course of writing this book. In particular, I should like to thank Felix Driver for his persistent enthusiasm, thoughtful comments and encouragement over a number of years. I am also endebted to Christopher Pinney, who took the trouble to read two versions of the manuscript and has always been ready with generous suggestions. Patrick Brantlinger, David N. Livingstone and Roy Porter each deserve my thanks for their insights and suggestions on earlier versions. I have learned much from my regular long-distance correspondence with Joan M. Schwartz and am grateful for her encouragement.

The research process would have been far more laborious and less interesting were it not for the many librarians, picture librarians and archivists who have gone out of their way to help me in my inquiries. I should like to thank especially Joanna Scadden of the Royal Geographical Society, London; Kathryn Hutton of the Foreign and Commonwealth Office, London; John Falconer of the India Office Library, London; Terry Barringer of the Royal Commonwealth Society Collections, University of Cambridge; Ann Datta of the Natural History Museum, London; Chris Wright of the Royal Anthropological Institute, London; Linda Atkinson and Sue Bird of the School of Geography Library, Oxford; William Schupbach of the Wellcome Institute for the History of Medicine, London; and Diana Madden of the Brenthurst Library, Johannesburg. I am grateful to Mrs Daphne Foskett for her kind permission to reproduce photographs by John Kirk from her private collection and to Russell E. Train for allowing me to examine a copy of Thomas Baines's storekeeper's notebook.

My colleagues at the School of Geography and St Hugh's College, Oxford, have provided a stimulating atmosphere for research and a friendly environment in which to work. In particular, I thank Elizabeth Baigent, Gordon Clark, Barbara Kennedy, John Wilkinson and the participants in the Postgraduate Geography Seminar. Martin Barfoot has cheerfully done fine photographic work, often to unreasonable deadlines.

The participants in the Imperial History Seminar and the London Group of Historical Geographers, both at the Institute of Historical Research,

London, have been especially helpful over the years in providing a friendly and scholarly atmosphere for research.

Various individuals have supplied thoughts, references and encouragement. Of those I can recall, I thank Kay Anderson, Tim Barringer, Andrew Blake, Chris Breward, Andreas Broeckmann, Kelly Boyd, Don Chapman, Richard Drayton, Elizabeth Edwards, Patricia Fara, David Gilbert, Peter Hansen, Sarah Harper, Jens Jäger, Phil Kinsman, Avril Maddrell Mander, Peter Marshall, Tim Meldrum, John MacKenzie, Rohan MacWilliam, Catherine Nash, Sue Parnell, Daniel Piercey, Andrew Roberts, Sarah Radcliffe, Jane Sampson, Raphael Samuel, Heather Shore, James Sidaway, Donald Simpson, Anthony Stockwell, Nicola Thomas, Adrianne Tooke, Gearóid Ó Tuathail and Bill Tunstall.

My extended family and friends have given much support for this project over several years. My biggest debt of gratitude is to my wife, Deborah, perhaps both my kindest and my most critical reader, for her astute comments and endless encouragement. Finally, it is unlikely that I would ever have been drawn to the cultural complexities of Empire were it not for the colonial experiences of my grandparents, and I thank in particular Arthur Horner for sharing with me his memories and insights.

Photographic Acknowledgements

Royal Commonwealth Society Collection, by permission of the Syndics of Cambridge University Library: 14, 20, 36, 80, 81, 83, 84, 86; By kind permission of Mrs Daphne Foskett (National Library of Scotland, Edinburgh: 10, 11, 12; The British Library, London: 38, 56, 57; Foreign & Commonwealth Office collection, London: 26, 39, 54; Courtesy of the Director, National Army Museum, London: 33; By courtesy of the National Portrait Gallery, London: 2, 3; The Natural History Museum, London: 87; By kind permission of the Royal Anthropological Institute, London: 4; Royal Botanic Gardens, Kew, London: 13; By kind permission of the Royal Geographical Society, London: 7, 9, 15, 16, 17, 18, 19, 20, 21, 22, 23, 24, 25, 27, 28, 29, 30, 31, 32, 34, 35, 27, 40, 41, 42, 49, 50 (also used on cover), 51, 52, 59, 60, 61, 62, 63, 64, 65, 66, 88, 91, and the Bodleian Library, University of Oxford: 4, 5, 8, 69, 70, 73, 90 and 92.

Abbreviations

AI	Anthropological Institute
BAAS	British Association for the Advancement of Science
BJP	*British Journal of Photography*
CO	Colonial Office
COVIC	Colonial Office Visual Instruction Committee
ESL	Ethnological Society of London
FCOL	Foreign and Commonwealth Office Library
FO	Foreign Office
GJ	*Geographical Journal*
GT	*Geographical Teacher*
IOL	India Office Library
JAI	*Journal of the Anthropological Institute*
JES	*Journal of the Ethnological Society*
JHG	*Journal of Historical Geography*
JRGS	*Journal of the Royal Geographical Society*
NAM	National Army Museum
NLS	National Library of Scotland
PRCI	*Proceedings of the Royal Colonial Institute*
PRGS	*Proceedings of the Royal Geographical Society*
RAI	Royal Anthropological Institute
RCS	Royal Commonwealth Society Collections
RGS	Royal Geographical Society
RHLO	Rhodes House Library, Oxford
SGO	School of Geography, Oxford
TESL	*Transactions of the Ethnological Society of London*
TIBG	*Transactions of the Institute of British Geographers*
UCL	University College, London
WI	Wellcome Institute

Introduction

When my great-great-grandmother posed for her picture to be taken on a postcard with an exotically attired rickshaw-puller in Durban, Natal, in August 1931, she was following a well-established pictorial tradition among European settlers and tourists in South Africa (illus. 1). Following the mass craze for picture postcards from around the turn of the century, such photographs became cheap to have made and post to family and friends back home as, in this case, fabricated evidence of the splendour of colonial life. Yet even before my distant relation had arrived to live in South Africa as a young bride in the mid-1880s, earlier generations of British settlers had had themselves immortalized in photographs in a similar fashion. Well before the picture postcard multiplied such images on a mass scale, photographs of white men and women being transported in rickshaws, palanquins and carrying chairs across all kinds of colonial terrain by 'native carriers' were common icons of the style of the British Empire. Similarly, the difference and lure of Africa had long been represented in visual imagery by the figure of an exotically dressed black man, as in well-known paintings such as Thomas Jones Barker's *The Secret of England's Greatness* (1861) (illus. 2), which depicts a colourfully dressed, bejewelled, anonymous black African king stooping to receive the Bible from a young Queen Victoria. In other words, by the time my great-great-grandmother chose, or was persuaded, to have herself photographed in this style, conventions of representing Empire were deeply set in the British imagination.

My book is concerned with the place of photography in this imaginative geography, from the start of Queen Victoria's reign (1837) to the eve of the First World War. In this period the invention and development of photography (from 1839) concurred with the extraordinary expansion of Britain's overseas Empire. At its zenith around the turn of the century the British Empire encompassed some 12 million square miles and almost a quarter of the world's population.[1] An indispensable record of the progress and achievements of Empire was provided by photography, which by 1900 could be as simple as pressing the button on a mass-produced Kodak camera. Whether drawn from family albums or official archives, historical photographs offer

1 'The author's great-great-grandmother with a rickshaw-puller, Durban, Natal, August 27th, 1931'.

compelling views of Britain's imperial past. Moreover, in the context of post-colonialism it is apparent that the photographic archive also represents a form of collective colonial memory, the making and longevity of which are only just beginning to receive the critical attention they deserve. It is against this background that I set out to consider the complex ways in which Empire was represented in photographs in the Victorian and Edwardian eras, as well as the role of imperial concerns in the practice of photography within a range of cultural domains.

In my use of the term 'imperialism' to describe the process by which Empire was fashioned, maintained and extended I follow much recent work by historians and post-colonial theorists which has argued that imperialism involved not only territorial acquisition, political ambition and economic interests but also cultural formations, attitudes, beliefs and practices. In particular, I support the contention of historians such as John MacKenzie who have described imperialism as a pervasive and persistent set of cultural attitudes towards the rest of the world informed to varying degrees by militarism, patriotism, a belief in

2 Thomas Jones Barker, 'The Secret of England's Greatness', 1861.

racial superiority and loyalty to a 'civilizing mission'.[2] Conceived thus, imperialism played a central ideological role within British culture and society in the Victorian and Edwardian eras, finding expression and nourishment in a range of cultural forms, including music hall, theatre, cinema, education, juvenile literature, sport and exhibitions.[3] It is only by considering photography within such a framework of cultural processes that it is possible to gain an understanding of the significance of the medium within British imperialism.

Following on from this, it is my contention that imperialism found sustenance in various photographic practices. The colonial photograph from my own family album already mentioned, for instance, represents relations between a white British woman and a black African man in ways unremarkable to a colonial settler such as my great-great-grandmother, and even perhaps to certain viewers today. However, one would surely find the picture surprising if the positions of the two figures were reversed. For whilst both are dressed in their finery with their faces on a horizontal level, they are clearly divided within the visual space of the image. What is more, the visual here concurs with the ideological: under close inspection the side of the rickshaw reveals the words 'The Durban rickshaw company. Europeans only'. As I will go on to show, photographic practices and aesthetics also express and articulate ideologies of imperialism.

Queen Victoria's iconic place at the hub of Empire was also projected

3 Robert Hills and John H. Saunders, 'Queen Victoria working at her dispatch boxes at Frogmore, attended by Sheik Chidda', 1893.

through photographs, such as one of her working at her dispatch boxes taken by the professional photographers Hills and Saunders (illus. 3). Here a composed Queen, dressed in her characteristic black 'widow's weeds', is shown dealing efficiently with the burden of ruling over the world's largest Empire. Her role as Empress of India is also given prominence by the upright figure of one of her Indian servants, Sheik Chidda, who is pictured patiently awaiting the royal command. Taken outdoors at Frogmore, Windsor, in 1893 on one of the Queen's regular visits to the Prince Consort's mausoleum, Hills and Saunders's photograph might have been taken in Simla or Delhi. Queen Victoria never visited India, yet such photographs, like the Indian-style Durbar Room at Osborne House, completed the same year, imaginatively transported one to the British Raj in India.

Although seemingly oblivious to the camera, Victoria was certainly not caught unawares, and maintained a fixed pose throughout the exposure; while the wind-ruffled tablecloth was blurred, her pen has remained still. The Queen understood not only the technical procedures of photography - indeed she was even more frequently photographed than painted - but also the important moral impact its imagery could have.

4 'Celebration of Queen Victoria's Jubilee, 1886', *Punch*.

Both the Queen and Prince Albert had been early enthusiasts of photography. They bought daguerreotypes as early as 1840, attended the first exhibition of the London Photographic Society in 1854 and even set up their own dark-room at Windsor Castle. An eager collector of family photographs, Queen Victoria also used the medium to great effect in the invention of the tradition of the Royal Family.[4] Photographs of the Queen and her family were available to the general public in mass-produced *carte-de-visite* portraits. Consisting of a print pasted on a small visiting card, these were used largely for portraits, notably of celebrities and 'exotic types'. One example was the 'Royal Album' produced by J. E. Mayall in 1861, which helped start the craze for this form of photographic ephemera. In due course, large prints and engravings of Queen Victoria came to be displayed in drawing-rooms, school halls and railway stations throughout the Empire.

In many respects, Britain's Empire, like much in the Victorian age, had the atmosphere and aesthetic charge of a grand spectacle.[5] This was perhaps at its most dramatic at the various exhibitions, jubilees and other popular collective celebrations of Empire. For example, *Punch*'s celebration of Queen Victoria's Jubilee of 1886 (illus. 4) conjures up in caricature something of the fervour of feeling at such displays of imperial might and progress. At the centre of the spectacle of Empire, portrayed here as Britannia, was Queen Victoria herself. While her name was bestowed on everything imperial, from waterfalls to cities, her figure was immortalized in countless sculptures adorning imperial cities from Bombay to Victoria.

The *Punch* celebration shows not only the spectacular pomp and ceremony associated with Empire, but also the figure of the photographer as both part and recorder of the scene. Photography became a vital ingredient in and witness of the Victorian culture of spectacle.[6] Surrounded by other inventions of the age, the photographer's camera is arranged and directed towards the distant horizon, mirroring the large cannon on the opposite side of this panorama of achievements of the age. A. R. Russell, author of *The Wonderful Century*, noted how the application of the sciences of chemistry and optics to the making of pictures 'belongs wholly to our time'.[7] As one of the key wonders of the Victorian age photography was thus widely regarded as a most powerful means of revealing the realities of the world and Britain's expanding presence in it.

'The Eye of History'

Many chroniclers of Britain's imperial story have drawn on its vast visual archive for illustrations of characters and events. Photographs, in particular, have been used as windows on the imperial past in moods of both nostalgia and critique.[8] However, many illustrated histories fail to place their images in anything but the most general of historical settings.[9] This stems from a more general historical myopia in using visual sources, particularly photographs which, as Raphael Samuel has noted, are treated simply as – and here he borrows a term from the nineteenth-century American photographer Matthew Brady – the 'eye of history', allowing us literally to look back upon the past:

> It is a curious fact that historians, who are normally so pernickety about the evidential status of their documents, are content to take photographs on trust and to treat them as transparent reflections of fact. We may caption them to bring out what – for our purposes – is tell-tale detail; but we do not feel obliged to question, or for that matter to corroborate, the picture's authenticity, to inquire into its provenance, or to speculate on why some figures are there and others ... are not. We do not even follow the elementary rules of our trade, such as asking the name of the photographer, the circumstances in which the picture was taken, or its date ...[10]

Samuel offers an important challenge: to put, as it were, quotation marks around historical photographs; to place them in their historical context in order to understand something of the imaginative worlds and cultural conventions which shape their meaning. As he also notes:

> The power of these pictures is the reverse of what they seem. We may think we are going to them for knowledge about the past, but it is the knowledge we bring to them which makes them historically significant, transforming a more or less chance residue of the past into a precious icon.[11]

Samuel is of course correct to argue that photographs are ambiguous images, since a multiplicity of meanings may be generated by the knowledge invested in them by their viewers. And yet it is precisely the chance residues of the past preserved in photographs – often in forms and details unforeseen by their makers – that give them their singular power as historical artefacts.

In looking at a painting, sketch or map we usually regard the object itself, noting its texture, material and optical patterns, as well as the subject it represents. When we look at a photograph, however, it is not generally the photograph we see but the referent itself. As Roland Barthes notes, 'Show your photographs to someone – he will immediately show you his: "Look this is my brother; this is me as a child," etc.'[12] The powerful visual presence conjured by a photograph stems from its status as an indexical sign, where the chemical tracings of light and dark are linked insolubly to a pre-photographic referent. This gives it a unique communicative presence; it becomes a selective but certain view of 'what has been' with a forensic charge akin to a fingerprint or death mask.[13] As Roland Barthes again notes, 'What the photograph reproduces to infinity has occurred only once: the photograph mechanically repeats what could never be repeated existentially.'[14]

To many Victorians, photography seemed to be a perfect marriage between science and art: a mechanical means of allowing nature to copy herself with total accuracy and intricate exactitude. Thus in 1854 at the twenty-fourth meeting of the British Association for the Advancement of Science (BAAS) Samuel Highley was reported to have

> demonstrated by a series of photographs what valuable aid might be rendered where *truthful* delineations of natural objects were of importance, as on disputed points of *observation* and how by the application of stereoscopic principles the student might in his closet study the Flora and Fauna of distant lands, or rare cases in medical experience.[15]

This faith in the ability of photography to render 'truthful delineations of natural objects' was behind many such calls for the application of photography to science and art from the middle of the nineteenth century. Moreover, as I will show in subsequent chapters, it was also this belief in the exactitude of the camera that shaped its uses in the exploration and survey of the peoples and landscapes of distant lands.

However, despite the common assumption that it was a truthful means of representing the world, photography was also a social practice whose meanings were structured through cultural codes and conventions. Indeed, photography could not have emerged as a 'modern point of view' without the pictorial conventions of perspectival realism that it inherited via landscape painting.[16] While photography certainly inherited many aesthetic conventions from painting, the indexical status of the photographic image noted above, together with the inability of the photographer to control completely the framing of the world by the camera – as opposed to the far greater degree of control exercised by a painter – made photography recognizably different from painting and all other graphic arts.

The insistence that photography's cultural currency is conventional, albeit to a lesser degree than that of painting or cartography, offers an important critique of much writing on the history of photography which has conventionally been dominated by a focus on the technical evolution of photographic processes and the artistic achievement of famous photographers. This history often begins with the 'invention' of photography in 1839 and culminates in photography's present-day self-consciousness.[17] Such an approach tends to assume that photography has some natural, fixed identity based on its technical evolution and aesthetic perfection. Traditionally photographs have also been treated as more or less transparent records of visual reality – 'windows on the world' which allow a complete and objective view into different times and places.[18] A predominant focus on individual photographers and their artistic production also tends, perhaps paradoxically, to read photographs as expressions of the unique vision of creative individuals. This approach not only prioritizes the aesthetic qualities of photographs but also marginalizes those practices of photography – from advertising imagery to scientific photography – that cannot easily be accommodated within a conventional art-historical framework.

Traditional models of photographic history thus largely fail to show how the authority ascribed to photography as transparent historical evidence of the past was itself constructed in particular historical contexts. However, the well-established emphasis on photography's technical evolution and artistic achievements has been broadened by studies which adopt social and cultural perspectives.[19] Just as visual perception in general is an active process, involving the continual extrapolation of information and the testing of theories on the basis of learned schema and cultural codes, photographs are selective, partial and legible within specific cultural frameworks.[20] As Allan Sekula has pointed out, 'If we accept the fundamental premise that information is the outcome of a culturally determined relationship, then we can no longer ascribe an intrinsic or universal meaning to the photographic image.'[21]

In a more extreme version of this argument the very coherence of a 'history of photography' has been challenged by critics like John Tagg who argue that

> photography as such has no identity. Its status as technology varies with the power relations which invest it. Its nature as a practice depends on the agents and institutions which set it to work . . . and its products are legible and meaningful only within the particular currencies they have.[22]

While I do not subscribe to Tagg's rather totalizing argument that photography is always and only a construction of discursive power, it is useful to emphasize that assumptions about photography's 'natural' evidential authority do need to be modified by a consideration of the history of what constitutes evidence, for if the photographic image can be interpreted as an analogue of visual perception, it is not because the former is a 'natural' transcription, but because both are historically and culturally coded.[23] Indeed, the photographic image derives its evidential force not only from its existential connection to a prior reality but from the technical and cultural processes and discursive frameworks through which it is made meaningful.

In the twentieth century, an ever more ubiquitous photography has itself become a model of human perception, the norm for the appearance of things. However, the analogy between the photographic image and the 'real world' is a conventional one; the camera's design stems from specific cultural conceptions of pictorially representing space in flat images, rather than from the structure of the human eye.[24] Yet, the camera-eye analogy helps conceal the highly pervasive assumption that the 'normal' mode of experiencing our surroundings is as a spectator looking at pictures. Knowledge is thus presumed to be a relation of 'correspondence' between a self-contained world on the one hand and a mental representation on the other.[25] Yet the world does not display itself to the observer. Representation is a complex cultural process and therefore photographs must be understood as moments in broader discourses, or 'ways of seeing', which require historical delineation.

An approach to photography as a culturally constructed 'way of seeing' also usefully emphasizes the intertextuality of photographic discourse. For photographs are not exclusively *visual* images. The 'common-sense' distinctions made between types of symbolic codes, including writing, photography, cartography and painting, are themselves historical.[26] Just as the Victorian novel employed forms of verbal visualization, framing devices, symbolic imagery and forms of perspective, so too photographs were seen and read in a complex interplay with other symbolic codes.[27] The meaning of photographic imagery is frequently framed by linguistic messages in the forms of titles, captions and accompanying text.[28] Moreover, photography has become a central practice in modern Western memory precisely because of the ways it has interacted with other kinds of texts.[29]

My arguments are thus oriented towards the construction of a 'visual history', charting the territory between approaches which on the one hand fail to situate images within their wider historical and cultural settings and on the other omit to examine critically the images themselves.[30] My starting point is that photographic images do not simply 'speak for themselves' or show us the world through an innocent historical eye.[31] Rather, they are invested with meanings framed by and produced within specific cultural conditions and historical circumstances. I thus argue that photographs – composed, reproduced, circulated and arranged for consumption within particular social circles in Britain – reveal as much about the imaginative landscapes of imperial culture

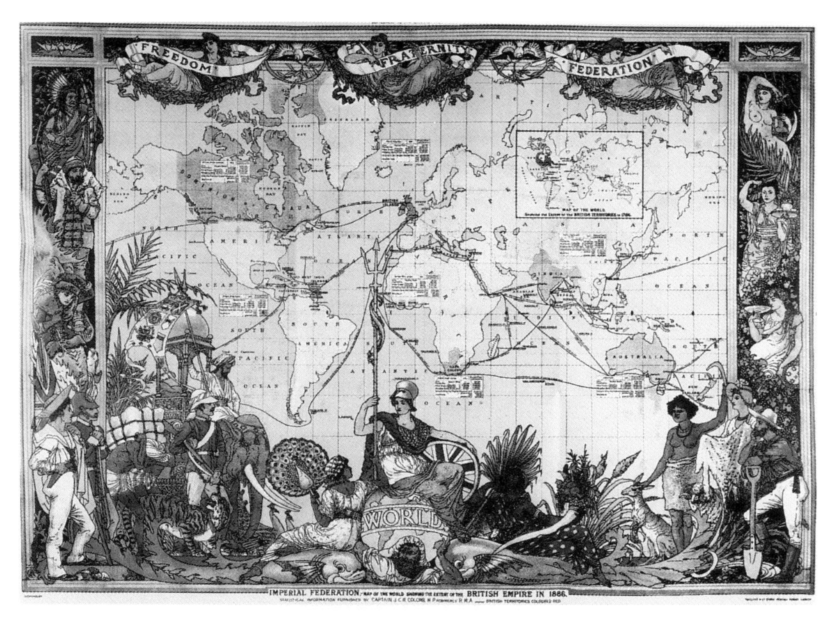

5 'Imperial Federation Map of the World', *The Graphic* (1886).

as they do about the physical spaces or people pictured within their frame. In this respect they are themselves expressions of the knowledge and power that shaped the reality of Empire.

Imaginative Geography

The British Empire was constructed in the Victorian imagination through a variety of cultural texts. One of its most enduring icons, projected onto the memories of generations of schoolchildren, was the world map with the territory of Empire coloured pink. The 'Imperial Federation Map of the World' (illus. 5), for example, was one of many such popular world maps produced towards the end of the nineteenth century to promote imperial unity. Published as a colour supplement to *The Graphic* to coincide with Queen Victoria's Jubilee of 1886, the map used Mercator's projection – which tended to exaggerate the relative size of Britain – and charted the growth of Empire over 100 years. Ironically, had the map been made just ten years later much of Africa would have also been coloured in British imperial pink. The world is framed by emblems of flora, fauna, racial types and colonial citizens from around the Empire, all of which are focused on the figure of Britannia, seated on the globe. The Imperial Federation was also behind the 'Howard Vincent Map of the British Empire', published by T. B. Johnston the same year. Measuring almost six foot square, this brightly coloured map, like the one reproduced in *The Graphic*, showed British imperial possessions in 1886, as well as an inset chart depicting the Empire in 1786. With additional information on population, revenue, steamboat routes and Royal Navy stations, it was clearly designed, as contemporary advertisements declared, 'to show the British People, in all Public Institutions, Free Libraries, and Schools, what the BRITISH EMPIRE is, and its great commercial value to the MOTHER COUNTRY'. The Imperial Federation maps, with their visual and textual promotion of imperial unity, proved popular and a number of subsequent editions were produced, including one in 1902 that showed an engraving of King Edward VII with Cecil Rhodes's declaration 'Union is Strength'. These maps were an important way of representing and promulgating a global geography from a British imperial perspective.

While the symbolic and practical importance of maps to imperial rule has been addressed by historians of Empire, cartography and geography,[32] the role of photography as a form of geographical discourse has been given far less critical attention.[33] Yet photography and cartography shared the ability to depict – to borrow the words from a popular world map and travel game of 1851 – 'geography made easy'. Like maps, photographs reduced the world to two dimensions. Indeed, photography was initially noted in circles of geographical science for its potential as a technique to assist cartography. It was in this context that George Greenough, the President of the Royal Geographical Society (RGS), noted in 1841, 'If one art more than another conveys to the mind a perception of the ideal . . . surely it is photography.' He praised photography's 'minuteness' and 'exactitude' as well as its seeming ability to act 'on the impulse of the moment, and with unerring certainty'.[34] Likewise Sir Roderick Murchison, during his second term in office as President of the RGS (1856–8), encouraged the use of photography as an accurate and economical means of reproducing maps.[35] Photography thus shared scientific geography's two-dimensional world view, and its ability to act 'on the impulse of the moment, and with unerring certainty' made it a significant means of legitimating geography's 'claim to reduce the world accurately to a uniform projection'.[36] Thus regularized and miniaturized, the world could be imaginatively seen, explored and possessed. 'The noblest function of photography,' one anonymous reviewer claimed in 1864, was

> to remove from the paths of science . . . the impediments of space and of time, and to bring the intellects of civilized lands to bear upon the phenomena of the vast portion of the earth whose civilization has either not begun, or is passing away . . . For the purposes of science, an explorer and a photographer should be convertible terms.[37]

Such faith in the ability of photography to expose the world to the scrutiny of science was shared by men of science and art. In fact, these were far from exclusive categories where, to coin a common contemporary expression, the

'ArtScience' of photography was concerned.[38] In particular, photography was implicated in a discourse of geography and a developing geographical science which took as its *raison d'être* the exploration and conquest of space.

My account of the imperial dimensions of photography is thus also offered as a contribution to a more contextual history of geographical knowledge, particularly in Britain's age of Empire.[39] For the growth of Empire coincided with the institutional development of geography - indeed, strong connections existed between the two.[40] As the President of the RGS noted in his anniversary address in 1860, 'With an empire that extends to every quarter of the globe ... the English have, perhaps, more to gain from the prosecution of geographical science than any other nation.'[41]

Founded in 1830 and incorporating the Society for the Exploration of Africa the following year, the RGS played an influential role in imperial society through its organizing of expeditions and its position as the focus of geographical knowledge. At once an academic archive, scientific club and imperial museum, the RGS was throughout the nineteenth century the institutional home of what one historian of geography has recently described as 'the science of imperialism *par excellence*'.[42] Led by champions of imperial science such as Sir Roderick Murchison,[43] the RGS occupied a pivotal position between the British scientific establishment and the imperial government, and between the élite world of imperial concerns and the wider public sphere.[44]

The place of photography within such an institution therefore deserves special attention and raises important questions about the interplay of science, photography and Empire. For a start, scientific institutions such as the RGS have played a central role in conditioning the making, consumption and survival of the imperial photographic archive.[45] The simultaneous positioning of the RGS in the domains of science, journalism and colonial officialdom also made it an important venue for the display of photographs. Evening meetings, such as one photographed in 1902 at the theatre of the University of London, Burlington Gardens (illus. 6), where meetings were held from 1871 to 1920, were rarely without the use of lantern-slides, particularly from 1890, when the RGS purchased its own projector. Although a projector is visible at the back of the theatre in this photograph, members of the audience are also holding hand-maps. The reporter's table is visible in the left foreground. Photography became so necessary to the recounting of travellers' tales and the display of geographical knowledge that when a lecture theatre was duly built at Lowther Lodge, the main premises of the RGS, it was designed especially to accommodate a large double screen for projection. The RGS's *Journal* and *Proceedings* were also quick to make use of half-tone techniques to publish photographs from the 1880s.

As well as a dark-room and copying room, the RGS also introduced a

6 'A Meeting at Burlington Gardens', from Hugh R. Mill, *The Record of the Royal Geographical Society 1830–1930* (1930).

'photograph room' to house photographic exhibitions and its rapidly growing collection of negatives, prints and lantern-slides. From the 1860s the RGS had begun to collect photographs from various sources, later requesting them from its Fellows through notes in its *Journal*. By 1930 it housed over 75,000 photographic prints and some 26,000 lantern-slides. The construction of this archive, with material gathered from the photographic exploits of Fellows around the globe and catalogued by geographical area in shelves of leather-bound volumes and hundreds of drawers of negatives, prints and lantern-slides, amounts to an attempt literally to capture the world in photographs. For, as Susan Sontag has noted, 'to collect photographs is to collect the world'.[46] Just as imperial geography possessed a central cartographic ambition to explore and fill in all the blank spaces on the world map, photography was soon part of an ambitious collective enterprise of visual survey. As visual 'facts', photographs were ideal for such an enterprise and played a central role within the imperial archive, which Thomas Richards has analysed in terms of 'a fantasy of knowledge collected and united in the service of state and Empire', and which was constructed precisely through the global capture of knowledge by scientific institutions such as the RGS.[47]

Photography's imperial scope certainly ensured that it played a key role in the institutional framework of British geographical science. In turn the apparatus of scientific societies such as the RGS – from the cameras it loaned to explorers to the photographic exhibitions it organized – did much to co-ordinate the making and display of the imperial photographic archive. Activities such as the training sessions in photography offered to explorers and travellers from the mid-1880s show further how the process of photography exemplified the grammar of observation and depiction at the heart of geographical science's quest to expose the unknown. Indeed, Sir John Herschel (1792–1871), the man of science who coined the very term 'photography', noted in 1861 that perfect descriptive geography should 'exhibit a true and faithful picture, a sort of daguerreotype'.[48]

While Herschel had called for a photographic geography, the commercial photographer John Thomson (1837–1921), Official Instructor in Photography for the RGS from 1886, called for a 'geographical photography'. As he claimed in a lecture on photography and exploration to the BAAS in 1891:

> Where truth and all that is abiding are concerned, photography is absolutely trustworthy, and the work now being done is a forecast of a future of great usefulness in every branch of science. What would one not give to have photographs of the Pharaohs or the Caesars, of the travellers, and their observations, who supplied Ptolemy with his early record of the world, of Marco Polo, and the places and people he visited on his arduous journey? We are now making history and the sun picture supplies the means of passing down a record of what we are, and what we have achieved in this nineteenth century of our progress.[49]

The importance of photography in representing Britain's worldly achievements in an age of Empire was clear to Thomson, a professional photographer who travelled and photographed widely in China, Singapore, and Cambodia between 1862 and 1872. It is worth discussing his work here briefly as it is particularly germane to my broader concern with the place of photography within geographical discourse.

As I will argue throughout this book, much Victorian colonial photography, from travel and topography to natural history, was broadly about geography. Following his return to Britain in 1872, Thomson found that his extensive photographic expeditions provided him with ample material for his subsequent publications, which married reproductions of his photographs with written accounts of his travels.[50] His travels in China, Cyprus and elsewhere were represented as a project to produce a photographic inventory of overseas resources and prospects for a British market interested in colonial expansion. One photograph included in his massive *Illustrations of China and Its People* (1873–4), for example, shows an arrangement of the 'Fruits of China' (illus. 7). Like other contemporary photographs of exotic fruit, such as those made by the English photographer Roger Fenton in 1860, Thomson was drawing on a painterly tradition of still life. Images of exotic fruit in European art had also been long associated with the bounty of new worlds beyond the seas and the contemporary iconography of the British Empire frequently drew on visions of exotic fruit being offered to Britannia (see illus. 5). Thomson's photograph does much the same by picturing and describing the fruit in detail, from its botanical classification to its texture and taste, and by noting its potential for being 'introduced into the orchards of Europe'.[51]

7 John Thomson, 'The Fruits of China', *Illustrations of China and Its People* (1873–4).

Thomson's use of photography to provide scientific detail is not uncharacteristic: he presented himself as much as a scientific explorer as an artistic photographer. From his joining of the RGS in 1866 and the Ethnological Society of London (ESL) in 1867, Thomson consistently promoted the scientific application of photography. Indeed, it is in this context, as much as in contemporary aesthetic conventions, that we should see his own photographs and travel narratives. For him, 'the faithfulness of such pictures affords the nearest approach that can be made towards placing the reader actually before the scene which is represented'.[52] His preferred use of the 'wet-plate' collodion process produced finely detailed photographic images which secured their 'scientific' status for geographers and anthropologists.

The impact of Thomson's work within geography, both through his photographic travel publications and his activities as Official Instructor in Photography from 1886, goes some way to showing that while the science of geography – through skills of map-making, topographic survey, resource inventory and strategic planning – was a 'sternly practical pursuit',[53] it was also a highly imaginative practice. This can also be seen, for example, in the diverse cultural readings and popular expressions of Henry Morton Stanley's explorations into 'darkest Africa'.[54] Yet despite moves towards a greater appreciation of the significance of imagery within geographical discourse, there has been little critical exegesis of specifically visual geographical imagery in the context of imperialism.[55]

In an attempt to analyse the place of photography within geographical knowledge and imperial discourse, I have therefore borrowed the term 'imaginative geography', first used by Edward Said in his study *Orientalism* (1978), in which he showed how 'the Orient' was constructed by Europeans as a set of complex and contradictory ideas and images of the East.[56] Whilst I am not concerned with specifically Orientalist representation it is worth noting, despite Said's focus on literary texts, that photographs made by European travellers, explorers, commercial operators and scholars constituted an influential form of Orientalist discourse.[57] More generally, photography – with its detail, ubiquity and currency across a range of institutional sites – played a significant role within the construction of the imaginative geography of Empire, creating a parallel Empire within a range of discourses including science, art, commerce and government.

I employ the concept of 'discourse' as it allows a variety of historically situated practices, concepts and institutions often considered separately – from hunting to mountaineering or from the Colonial Office to the Boy Scout Movement – to be discussed critically in relation to each other. I have no wish to overplay the consistency and uniformity of cultural forms of imperial power.[58] Indeed, I believe firmly that the complexities of imperialism and colonialism can be delineated only through their various and specific expressions.[59] I have therefore not set out to produce some grand survey of imperial photography in all its guises. Rather, I aim to narrate a series of more localized photographic explorations of specific persons, places and practices, while nevertheless relating them to broader thematic trends and ideological frameworks. In so doing I wish to contribute to recent important work by archivists, historians and historians of anthropology who have begun to give the imperial photographic archive the critical historical attention it deserves.[60]

Photography was used to picture the British Empire in a variety of ways. In the chapters that follow I will attempt to identify a number of key themes in which Empire's imaginative geography was produced and projected through practices of photography. As a technology based on the power of light, photography assumed particular symbolic significance as part of geographical discourse informed by what has been aptly termed a 'providential theology of colonial praxis' whereby the mutual extension of Christian civilization and scientific knowledge represented a transference of 'light' into the 'dark' recesses of the globe.[61] This iconography was articulated in one of the major facets of nineteenth-century geographical discourse: exploration. I will begin, in Chapter 1, by considering the role of photography in the geographical exploration of Africa through David Livingstone's Zambezi Expedition of 1858–64.

Scientific explorers were not the only travellers who used photography and in Chapter 2 I will examine the topographical and landscape work of commercial photographers of the 1860s and 1870s, looking in particular at the travels of Samuel Bourne in India and John Thomson in China and Cyprus. As we will see, landscape photography, in the context of both science and art, was a powerful means of organizing and domesticating imperial landscapes.

In Chapter 3 I consider the place of photography in another important mode of exploration and colonial encounter: military campaigning. After considering the adoption of photography by the military, particularly the Royal Engineers, I will provide a detailed discussion of the Abyssinia Campaign of 1867–8. This campaign, which involved both official Royal Engineer photographers and a host of scientific explorers and journalists, reveals how closely photography and science were involved in the practice and representation of imperial warfare.

Contemporary accounts of military campaigning and exploration often alluded to one of the most spectacular practices associated with British colonial rule: hunting. In Chapter 4 I trace the language and imagery of 'hunting with the camera', particularly in Africa, noting its associations with a range of other practices and languages, including taxidermy, mountaineering and conservation.

In Chapter 5 I will explore how photography was used in the survey and classification of 'racial types'. In particular I situate the role of photography in the sciences of anthropology and geography within the context of Victorian theories on 'race', both in terms of non-European peoples in the overseas Empire and social groups within Britain, particularly those 'others' of Victorian London. I will consider how photographs mobilized, reinforced and contradicted shared understandings of cultural difference.

Photography was most effectively used as a vehicle for promoting imperial messages on a large scale through lantern-slide lectures. In Chapter 6 I will examine the use of lantern-slide lectures in the teaching of imperial geography to British schoolchildren. I will focus on the work of the Colonial Office Visual Instruction Committee (COVIC) from 1902 to 1911, looking in particular at the role of the educationalist and geographer Halford Mackinder and the project's official photographer, Hugh Fisher. As I will show, the aims and techniques adopted by the COVIC align it with a broad range of cultural currents of imperial teaching and promotion in the early twentieth century.

Finally, in Chapter 7, I will attempt to draw together some of the themes of the earlier chapters to consider the place of photography within the imaginative geography of Empire and to raise questions about the shifting status and mutable meaning of historical photographs.

1 Exploring Darkness

Almost immediately after the invention of photography in 1839 attempts were made to employ the new technique on various European overseas explorations. Sir John Herschel, for example, the eminent English astronomer, tried unsuccessfully to have photographic apparatus included on the British Antarctic Expedition of 1839.[1] In 1846 it was suggested in the journal *Art Union* that the 'Talbotype'

will be henceforth an indispensable accompaniment to all exploring expeditions. By taking sun-pictures of striking natural objects the explorer will be able to define his route with such accuracy as greatly to abbreviate the toils and diminish the dangers of those who may follow in his track.[2]

The Talbotype (or calotype), named after the negative photographic process invented by William Henry Fox Talbot in 1840, used sensitized paper negatives which, although they did not provide particularly sharp images, were easier to transport than glass. The technique, which remained popular until the 1850s, was thus seen by many as well suited to use by travellers and explorers. However, despite calls for the application of photography on journeys of exploration and the enthusiasm of distinguished scientists such as Sir John Herschel and Sir David Brewster,[3] the technical skill and bulky equipment necessary for early photography, together with preference for more established forms of depiction, hindered its immediate deployment on expeditions.

Until the mid-1850s, therefore, photography remained beyond the reach of most ordinary travellers. Improvements in photographic apparatus, particularly the development of Frederick Scott Archer's 'wet-plate' collodion process (1851), made expeditionary photography more commercially viable, but involved even more cumbersome apparatus and technical manipulation. The difficulties faced by the early expeditionary and travel photographer using this popular technique are evident from contemporary manuals and engravings such as 'Photography and Exploration' in Gaston Tissandier's *A History and Handbook of Photography* (1876) (illus. 8). Archer's process involved coating a glass plate evenly with collodion (a solution of nitro-cellulose

8 'Photography and Exploration', Gaston Tissandier, *A History and Handbook of Photography*, (1876).

and ether) containing potassium iodide, sensitizing it by immersing it in a basin of silver nitrate and then exposing it in the camera while still wet. Because plates had to be coated, sensitized, developed and fixed on the spot, photographers had to carry plates, chemicals and a dark-tent with them. As this engraving suggests, early photographic operations on expeditions were by no means straightforward. However, despite the transportation difficulties, the technique was taken up by most photographers because the process produced much finer images than calotypes, required an exposure time of a few seconds as opposed to a few minutes and was free from English patent restrictions, which continued to limit the use of Talbot's calotype until the mid-1850s.

By the 1860s photography was being employed on an ever larger scale by a range of British explorers heading towards the 'lesser known' regions of the globe. Just as there was a complex array of motives behind Victorian exploration, including romantic adventure, commercial prospecting, military conquest, geographical discovery and the pursuit of scientific knowledge, photography was deployed for a correspondingly wide range of purposes and effects. However, despite their rhetoric of adventure and discovery, explorers were seldom representing something entirely new. Rather, they were employing new ways of representing the unfamiliar.

Perhaps not surprisingly, many travellers and explorers had great difficulties with their photographic equipment, or lost their work because of heat, damp or breakage. James Grant, accompanying John Hanning Speke on the famous 1860–63 expedition to the source of the Nile, succeeded in making a number of stereoscopic photographs in Zanzibar at the start of the expedition, but subsequently found the camera too awkward and concentrated on producing coloured sketches instead.[4] It is thus not surprising that photography was applied unevenly in the wide-ranging expeditions that so marked the Victorian age. It was noticeably absent, for instance, on journeys to the interior of Australia until the late 1880s. One historian has claimed that since photography was thought to capture everything objectively and completely, explorers in Australia resisted it, thinking that it threatened to render their own, individual narratives invisible.[5] However, while the omission of photography on expeditions into the Australian interior until the late 1880s is striking, it appears largely exceptional. Conscious or subconscious resistance to the medium from explorers, save for the obvious practical and technical reasons, is not in evidence elsewhere, as explorations of the interior of 'darkest Africa' from the 1850s clearly demonstrate.

The image of Africa as 'the dark continent' was a powerful one to much of the British public in the second half of the nineteenth century. As a technology based on the natural power of light, the camera seemed particularly suited to the task of illuminating the secrets of the continent. However, through their supposed power to reveal the unknown and the geographical truth, photographs made by British explorers in Africa tended to reinforce the established image of the African interior as a place of disease, death and barbarism. By producing such imagery as the visual truth, photography played a significant part in the process by which, as Patrick Brantlinger has put it, 'Africa grew "dark" as Victorian explorers, missionaries and scientists flooded it with light'.[6]

The rhetoric of light and dark was particularly prevalent within the visual and verbal imagery of Christian missionary discourse. On John Gilbert's 'Pictorial Missionary Map of the World' (1861), for example, sold by James Nisbet & Co., London for a penny, much of Africa was coloured black, depicting 'heathen' territory, illuminated only by white missionary stations and rivers. More generally, the associations of whiteness with Christianity, civilization and European racial identity would not have been lost on viewers of engravings and paintings depicting Britain's colonial encounter with Africa, such as Thomas Jones Barker's *The Secret of England's Greatness* (see illus. 2). In the same period pictures of the missionary-explorer David Livingstone, one of the greatest imperial heroes of the Victorian age, depict him as a radiant white figure bathed in light amidst the dark and dangerous surroundings of tropical Africa.

This iconography of light and dark also shaped the practice of photography on expeditions to Africa. Indeed, David Livingstone's Zambezi Expedition of 1858–64 represents its earliest use on an official British expedition. Livingstone had long favoured optical technology in the form of a magic lantern to project the word of God to Africans. While illustrating stories from the Bible, it also served to demonstrate the superiority of European technology.

Indeed, Livingstone referred to the magic lantern, at least one of which was taken on the Zambezi Expedition,[7] as the 'oxyhydrogen light of civilization'.

The Zambezi Expedition evolved as part of Sir Roderick Murchison's masterminding of national enthusiasm for David Livingstone. During his second presidency of the RGS, Murchison recruited official sponsorship (to the tune of £5,000) from an otherwise hesitant government and ensured the RGS and the Royal Society played a central and public role in the expedition's organization and equipment. Through its role as an organizer of expeditions and a venue for the display of travellers' tales, the RGS was at the forefront of opening up distant scenes to the scrutiny of British scientists, travellers and an interested public. The interior of Africa had appeared to offer exciting prospects to Europeans from the late eighteenth century. From 1830 the RGS effectively took over the expeditionary thrust begun by the African Association, founded in 1788, and Africa was to remain the single most important destination for its expeditions throughout the nineteenth century.

Although the science of geography was pursued through a number of channels, overseas exploration represented the focal point of the RGS and, as such, lay at the heart of the Victorian geographical enterprise. The promotion and practice of expeditions were the main means the RGS used to consolidate itself as the institutional centre of geographical science and to endow geography with an aura of national usefulness. As the President put it in 1875, 'Exploration indeed is the pioneer of progress.'[8]

Although often presented as a simple story of topographical serendipity and heroic discovery, exploration in general and British exploration in the 'age of Empire' in particular were always animated by a sense of mission. By definition, exploration is both purposeful and seeking. The goals set out, the landscapes imagined and the forms of observation and categories of knowledge employed were more often than not framed in the explorers' own cultural terms. Furthermore, exploration not only influenced the shape of scientific institutions such as the RGS but had a profound effect on both popular attitudes and official policy towards those lands and peoples fast coming into the orbit of Britain's imperial influence.

The official aim of the Zambezi Expedition, as set out in David Livingstone's letters of instruction to his officers, was explicit enough:

> The main object of the Expedition . . . is to extend the knowledge already attained of the geography and mineral and agricultural resources of Eastern and Central Africa, to improve our acquaintance with the inhabitants, and engage them to apply their energies to industrial pursuits, and to the cultivation of their lands with a view to the production of the raw material to be exported to England in return for British manufactures.[9]

The wider purposes envisaged by Livingstone included the extension of 'legitimate' (non-slave based) commerce and Christianity, together marking the beginning of the civilization of Africa.

The scientific aims of the expedition demanded a range of skills and techniques of representation, including photography. It was in this capacity that David Livingstone took his brother Charles along as official photographer and cartographer. That photography was deemed important enough to merit an official place on the expedition is in itself significant. In a letter to Charles, written in May 1858, David Livingstone suggested that from the early stages of the expedition his brother should get his photographic apparatus working in order to 'secure characteristic specimens of the different tribes . . . specimens of remarkable trees, plants, grain or fruits and animals', as well as 'scenery'.[10] Thus the practice of photography was shaped clearly in accordance with the aims of the expedition and, despite the bulky apparatus necessary, was embraced as a powerful new means of recording permanently the landscapes, inhabitants, flora and fauna of the area to be explored.

Only a single stereoscopic photograph of a baobab tree made by Charles Livingstone on the Zambezi Expedition (now housed in the collection of the National Museum, Livingstone, Zambia) appears to have survived the test of time. Yet although Charles was notoriously inept at most of his expeditionary duties, his photographic efforts were far from failures and he returned home with some forty stereoscopic negatives. Late in 1863, after his return from the expedition, he wrote to the Foreign Office, claiming:

> I have been engaged in making arrangements for printing about 40 different stereoscopic photographs of the natives in their various occupations and amusements, some remarkable trees, rocks, etc. for the use of Sir Roderick Murchison and Professor Owen. The printing of these photographic specimens will cost about £6.[11]

Richard Owen, the famous naturalist and critic of Darwin, had little hesitation in supporting Charles's claim to the Foreign Office:

> With respect to the photographs, as these are most useful & faithful records of the physical characters of the native tribes, I suggested the desirability of their being printed, in the interest of Ethnology. I have no doubt that the photographs of rocks would thereby be made equally useful to the Geologist and of the trees to the Botanist . . . I am of the opinion that the services which Mr C. Livingstone has rendered in England are such as to call for remuneration.[12]

Following such a commendation, Charles was duly awarded an extension to his expedition salary and the cost of printing the photographs. It would certainly seem that he fulfilled his brother's initial ambitions: his photographs were seen to make a significant, scientific contribution to knowledge of the geography, ethnography, geology and botany of Africa. The publishers John Murray also used them in preparing the illustrations for David and Charles Livingstone's *Narrative*, claiming in the preface, 'Photographs by Charles Livingstone and Dr Kirk have materially assisted in the illustrations.'[13]

As this suggests, Charles was not the only photographer on the expedition.

John Kirk, the botanist and medical officer of the party, was an experienced amateur photographer and, taking his own equipment on the expedition, was quick to comment on Charles's early efforts, notably how he had 'made a mess of it'.[14] Kirk successfully made negatives using waxed paper as well as the collodion that Charles used.[15] Waxed-paper negatives, developed by Gustave Le Gray in 1850, gave finer-quality negatives than the normal calotype paper, but were not as detailed or as sensitive as collodion. Kirk also worked with Charles in the early part of the expedition, helping him with his photographic processes. In June and July 1858 Charles, Kirk and Thomas Baines, the expedition's artist and storekeeper, were left on Nyika Island while the rest of the expedition transported equipment upriver. As Baines recorded in his diary:

> Mr Livingstone got out his dark tent for photography and we set it up . . . In the afternoon the stereoscopic camera was set up and we grouped ourselves and the Kroomen in [and] about the house and shed and took half a dozen views, some of which Kirk succeeded in developing at night.[16]

Although almost all of the surviving photographs of the expedition are those made by John Kirk, both he and Charles Livingstone spent considerable time experimenting with photographic processes. Moreover, despite many initial failures, due largely to malaria, exhaustion and his inexperience with photographic apparatus, Charles claimed in a letter to his wife to have got some 'good negative pictures which I hope to sell in England', adding later that he managed to take a perfect picture of a baobab tree.[17] Charles's ambition to sell his prints back home betrays some of the personal commercial motivations existing within his ostensibly scientific purpose. He was clearly searching for saleable scenes of the picturesque and the unusual, as well as for scientific specimens. A month later, after returning from exploring and photographing upstream, Charles noted: 'Since my return I have been taking some prints – got 2 of women, one with their water pots on their heads & the other some in gala dresses.'[18]

As well as making photographs illustrating African life and manners, the expedition's photographers also surveyed important topographical features, particularly those relating to the navigation of the Zambezi. It was in this capacity that the camera was employed in the exploration and recording of the Kebrabasa rapids in November 1858. As the prime obstacle to David Livingstone's dreams of a navigable Zambezi through which he hoped to open up the interior of this part of Africa to civilization and 'legitimate commerce', the Kebrabasa rapids necessitated careful exploration and survey. He thus noted in his diary: 'On reaching the place we have as yet called Kebrabasa . . . Baines sketched whilst Mr. L.(Charles) photographed them.' Thomas Baines's painting *Shibadda, or two channel rapid, above the Kebrabasa, Zambe-*

9 Thomas Baines, *Shibadda, or two channel rapid, above the Kebrabasa, Zambezi River*, 1859.

zi River (illus. 9), worked up from a sketch of a scene above the Kebrabasa Rapids, shows a camera and tripod on an outcrop of rock with Charles Livingstone working under its black cloth and what is likely to be either David Livingstone or John Kirk standing beside him.

Charles certainly took his camera to the rapids. His journal entry for 25 November reads:

> got my photog., placed it on a rock & took a view. It was with a dry prepared plate & took half an hour after I got all adjusted. I was all rather "used up" before by the long walk but this roasting in the sun did for me.[19]

Despite his bodily sufferings, Charles seems to have secured a good negative. His protracted preparations also fixed an image in Thomas Baines's visual unconscious. The art historian Tim Barringer has read Baines's depiction of photography here as one of the experienced artist mocking a threatening new technology.[20] Indeed, Baines includes himself in the foreground, holding his completed sketch and enjoying a drink, while he depicts the photographer and assistant stuck on a rock in the distance, rather dwarfed by the surroundings, attempting to capture the rapids. However, I am not convinced by this interpretation, for while Baines did not have the best relationship with Charles, he was genuinely admiring of the art and science of photography. Moreover, although their equipment was different and the resulting images held different currencies of veracity, the roles of expeditionary artist and photographer were closely associated. Thus David Livingstone had also instructed Thomas Baines to make 'faithful representations of the general features of the country through which we shall pass': sketches 'characteristic of the scenery'; 'drawings of wild animals and birds'; pictures of 'specimens of useful and rare plants, fossils and reptiles'; drawings of 'average specimens of the different tribes ... for the purposes of Ethnology'.[21] Painting was thus, like photography, intended to be applied as a means of capturing geographical and ethnological knowledge. Moreover, after his dismissal from the Zambezi Expedition on what now seem highly spurious charges of stealing from the stores, Baines teamed up with the trader, hunter and photographer James Chapman in an expedition to the Victoria Falls during which the glass plate and the sketch pad were used in even closer conjunction. In subsequent years the camera came to figure in a number of Baines's African sketches and paintings, less in a spirit of competition than in one of shared ambitions for both scientific accuracy and artistic attraction. Indeed, I think it is more helpful to regard the camera in Baines's paintings as representing the artist's attempt to stamp his pictures with the authority of a photograph. His painting of the Kebrabasa Rapids thus emphasizes how, far from being rival techniques, photography and painting were closely associated in the making of records of the geographical features and prospects of the country.

In this respect expeditionary photography was reprocessing a longer tradition of making picturesque 'views' within scientific imperial exploration and survey. In his *Picturesque Views on the River Niger* (1840), for example, Commander William Allen presented views 'made on the spot' during Lander's unsuccessful Niger Expedition of 1832–3, which he accompanied as commander of HMS *Wilberforce* and surveyor of the river. Both Allen's views of the Niger and the photographs, paintings and maps of the Zambezi were made as part of official expeditions closely associated with the RGS and the British Government – which aimed to combat the slave trade and establish Christianity and 'legitimate commerce' in Africa. *Picturesque Views*, described by Allen as an 'endeavour to delineate the features of the country, and the manners of the people', itself represented a geographical survey in which the view and the map were read together. Thus on the map of the 'Rivers Niger and Chadda, Surveyed in 1832–3 by Commander W. Allen' crosses marked the spot from where the views were taken. Through the sequence of numbered views and accompanying captions, Allen guided the reader and viewer on an exploration up the river.[22]

With their indexical currency photographs had even greater potential use within such survey work. It is for this reason that David Livingstone included them, along with maps and sketches, with his written expedition report of December 1858 to the new Foreign Secretary, Lord Malmesbury, which was later exhibited at evening meetings of the RGS.[23] Describing the party's exploration and survey of the rapids, Livingstone expressed his conviction that a steamer could pass over them 'without difficulty' when the river was in flood. He backed up his assertion by noting that 'a careful sketch and photograph were made of the worst rapid we had then seen', adding that the former referred to the 'water colour by Mr Baines' and the latter to the 'photograph by Mr C. Livingstone'.[24] He explained that he had included visual images with his despatches because 'I thought this the best way of conveying a clear idea of my meaning'. Livingstone noted in particular a photograph 'showing a dead hippopotamus while also exhibiting the rock in the river' and 'another photograph [which] exhibits the channel among the rocks'.[25] He refused to admit that the Kebrabasa Rapids could thwart his plans for a navigable Zambezi, even after he and Kirk, who knew better, had later encountered even larger rapids further upstream. Thus, far from showing what Kirk knew to be true and Livingstone refused to believe, the visual evidence was designed to reinforce Livingstone's assertion that a suitable steam vessel could navigate the entire length of the river. Informing the Foreign Office of the absolute obstacle to navigation posed by the Kebrabasa Rapids would have thrown uncertainty over not just Livingstone's request for a new steamship but over the entire future of the expedition.[26] He thus used photographs – stamped with his own interpretation – as a form of incontrovertible evidence, a means of naturalizing his colonial vision: rendering the unfamiliar familiar and the unknown known; converting complex environments into the constituent categories of European scientific knowledge.

Other photographs and sketches of the Zambezi and Shire rivers were also made by John Kirk.[27] Such views effectively map these rivers and their obstacles to navigation. The convincing evidence seemingly produced by photographs also made them particularly suitable witnesses to the discovery and naming of 'new' geographical features. Kirk photographed a section of rapids on the Shire which he and David Livingstone claimed to have discovered on 9 January 1859 (illus. 10).[28] They named these rapids, above which the expedition ships could not navigate, Murchison Rapids, after Sir Roderick Murchison. They were neither the first nor the last to so honour this man of science. Indeed, Murchison Rapids was one of five geographical features in Africa and twenty-three worldwide which would come to carry his name. Such naming mapped Murchison's significance within metropolitan scientific and political circles, marking his pivotal role in promoting exploration in Africa and beyond.[29] In the process of the 'scramble for Africa', the names of famous scientists, monarchs and explorers – the latter most incisively portrayed in a 'Map of African Literature' in William Winwood Reade's *An*

10 John Kirk, 'Murchison Rapids, River Shire, 1859'.

African Sketchbook (1873) – were scattered liberally across Africa, recasting mountains, rivers, valleys and even whole countries in the nomenclature of European authority. By simultaneously photographing and naming the Murchison Rapids, Kirk and Livingstone were commemorating their sponsor and inscribing, in their own terms, their geographical discovery on to the landscape. The power of photography as a historical witness was already rendering it an important vehicle for superimposing icons of British civilization and authority on to the African continent.

As I have established, photography was embraced on the Zambezi Expedition as a powerful new means of producing geographical knowledge about eastern and central Africa. The resulting photographic representations of Africa, like their cartographic and textual counterparts, were by no means neutral since they were part of an expedition whose expressed goals envisioned the establishment of European settlement and commerce. David Livingstone articulated this clearly in his introduction to the narrative of the Zambezi Expedition, published in 1865:

> In our exploration the chief object in view was not to discover objects of nine days' wonder, to gaze and be gazed at by barbarians; but to note the climate, the natural productions, the local diseases, the natives and their relation to the rest of the world: all which were observed with that peculiar interest, which, as regards the future, the first white man cannot but feel in a continent whose history is only just beginning.[30]

11 John Kirk, 'Lupata July 13th 1859'.

Photography became part of this ethos as it was an ideal means of representing, as visual 'truth', the potential and problems of British imperialism in Africa.

A number of the photographs made on the Zambezi Expedition also depict the visual iconography of 'darkest Africa'. Kirk's photograph 'Lupata July 13th 1859', for instance, represents a wall of impenetrable, twisted vegetation (illus. 11). It fits into a well-established image of Africa as a place of barbarism and savagery that was both reinforced and complicated as more of the continent was explored and visualized by Europeans. The image of 'the dark continent', spun together from various impressions circulating in Europe since its earliest contacts with West Africa in the fifteenth century, began to solidify in the mid-nineteenth century as Europe – particularly Britain – started seeing in Africa opportunities for market-based commerce and, by extension, more permanent imperial prospects. By the 1850s explorers left Britain equipped with common images and assumptions about the nature of Africa and the Africans' place in nature.[31]

The dense foliage of Africa's tropical environments, for example, had been regarded by generations of Europeans as offering both the promise of abundant riches and the dangers of the unknown, of disease and death.[32] Richard Burton's book *The Lake Regions of Central Africa: A Picture of Exploration*

13 Herbarium specimen of *Barringtonia racemosa (L.) Spreng,* collected by John Kirk, October 1862.

12 John Kirk, 'Senna. ? July 1859'.

(1860) painted, as its subtitle suggested, yet another gloss on this image. The then President of the RGS, Earl Grey, remarked that the landscape Burton described had a 'repulsive aspect', being 'a fever-stricken country that is skirted by a wide, low-lying belt of overwhelming vegetation, dank, monotonous, and gloomy, while it reeks with fetid miasma'.[33] Kirk's photograph represents this imaginary landscape: one that is not yet free of a state of barbaric, primeval geography.

As botanist and medical officer to the expedition, Kirk was particularly interested in both 'natural productions' and 'local diseases' as Livingstone had termed them. Extensive medical debates in Britain and West Africa had long focused on establishing the causative connections between types of disease and 'miasmas' (pockets of poisonous air) associated with damp, luxuriant and, importantly, uncultivated tropical environments. Kirk's choice of scene thus correlates with the concerns of mid-nineteenth-century medical topography and was framed by his interest in establishing the suitability of tropical Africa for development by the 'white races'.[34]

The application of photography to medical topography on colonial expeditions had been anticipated for a number of years. In 1846, *Art Union* urged the widespread use of Fox Talbot's apparatus within colonial exploration, noting:

> In the exploration of African rivers it has been found that some spots are fearfully infected by miasmata and malaria, while others at a little distance are safe and salubrious. Now, Talbotypes would obviously be better guides to these spots than the best written descriptions.[35]

John Kirk's practice of photography was situated within the context of such claims for the scientific application of photography as well as his own interests in medicine and botany.

Both Kirk and Charles Livingstone played a central role in amassing collections of flora and fauna for further study in England, and their photographs are clearly part of the expedition's broader concerns with collecting for natural history. The ringing endorsement given to Charles's collection of 'photographic specimens' by the naturalist Richard Owen provides further evidence of how seriously photography was taken by such men of science. As botanist to the expedition, Kirk was especially interested in applying photography as a means of recording specimens of flora. Indeed, the majority of his surviving photographs show trees and scenes of vegetation (illus. 12). Kirk's photographs of trees such as the baobab, or creepers, selected expressly to contribute to botanical discovery, have greater resonance as natural history specimens than as mere scenic views of African landscape. The extended album caption to Kirk's photograph 'Lupata July 13th 1859', for example, notes:

Vegetation at Lupata, selected to show the fleshy twiners which are so abundant there: their stems although five inches in diameter may be cut through at one cut with a knife, when a great quantity of milky juice runs out, this contains a good deal of caoutchouc.

Kirk sent photographs, sketches, written descriptions and plant specimens to the Royal Botanical Gardens at Kew, headquarters of Britain's empire of nature. While Kirk's photography was influenced by natural history collecting, his specimen hunting was also shaped by photography; many of his plant specimens, with their pressed leaves and stems mounted on paper, resemble early photographic leaf prints in their fine detail, forensic capacity and aesthetic arrangement (illus. 13). Nevertheless, it would appear that while photography was useful in recording the larger types of flora, particularly trees, it was not about to take over from the practice of collecting plant specimens, where colour and texture as well as seeds could be preserved. Such specimen hunting was certainly motivated in part by aesthetic concerns, particularly the search for interesting objects with which to illustrate travel narratives. Yet in contributing to Kew's storehouse of botanical knowledge, the collectors on the Zambezi Expedition were preoccupied by more practical considerations, notably the discovery of commercially significant natural resources. Samples of coal and wood were thus collected and their potential uses as fuel for the expedition steamship, the *Ma Robert*, and future steam vessels closely noted. John Kirk in particular paid much attention to potential productions from plants, from the extraction of rubber and the production of cotton to the making of baskets and nets from the baobab tree.

Unlike Charles Livingstone, Kirk does not seem to have been interested in making photographs of the indigenous inhabitants of the areas the expedition visited. Indeed, most of his surviving images present few signs of either the peoples of the region or the expedition of which he was part. This selective vision of topographical features and botanical specimens represents in part the exercise of that 'peculiar interest', as David Livingstone had put it, with which Europeans imagined Africa as a continent without history until colonization. This imaginative stricture was also applied to the indigenous inhabitants of 'the dark continent'. Kirk's photographs present an ostensibly empty landscape. Along with cartographic and literary representations, photography was selectively deployed within a geographical discourse which emptied lived environments of their human presence and in turn isolated indigenous peoples from their habitats.[36] Kirk's botanical specimens in this sense become the counterpart to Charles Livingstone's representations of ethnographic 'specimens'. In both cases the camera became a particularly effective tool of naturalization.

Although largely absent from Kirk's photographs, the inhabitants of the areas visited by the Zambezi Expedition were in conceptual terms central to the representation of 'savage' landscapes. As Francis Galton noted, accounts of British explorers like Burton projected 'a repulsive picture of a vulgar, boisterous and drunken savagery over-spreading the land'.[37] Like a number of his contemporaries, Galton shared Burton's unfavourable view of central Africa. Galton here followed Earl Grey, the President of the RGS, who noted in 1860 that Burton had projected 'the picture of one unbroken spread of vulgar, disunited, and drunken savagery over the entire land'.[38] The image of Africa's environment and inhabitants together constituting 'one rude chaos' was hardly new.[39] However, from the late eighteenth century Europeans had begun to consider the apparent fertility of the tropical environment as both cause and indicator of social idleness.[40] This was especially significant for those like Livingstone who believed that all humans had originated from a single source (monogenesis) and argued that environmental and human improvement were possible. Images of vegetative pandemonium such as that projected in Kirk's photograph (see illus. 11) signified not only the rich potential of the land but also the absence of indigenous industry and labour. Thus images like Kirk's reinscribed an imaginative geography of the 'dark continent'. While the African landscape was represented as disorderly, even threatening,[41] it was also presented as a colonial prospect, where wildness could be taken for unruly fertility and could be read as a blank space for improvement.

While the Zambezi Expedition was one of the earliest attempts to use photography on expeditions to Africa, it was not the only one. Indeed, despite the difficulties posed by harsh environmental conditions and bulky apparatus, an increasing number of explorers took up photography in order to record their discoveries. James Chapman, for example, took photographic apparatus on his hunting and trading expeditions in the interior of South Africa from 1859–63.[42] In 1859–60 Chapman accompanied Thomas Baines, after the latter had been dismissed from Livingstone's Zambezi Expedition, to the Victoria Falls on the Zambezi. In a letter of January 1860 to Sir George Grey, the Governor of Cape Colony, Chapman explained how problems of illness, drought and poor sport had been compounded by a greater failure: 'Of all our little disappointments I regret none more deeply, and I am sure your Excellency will sympathise with me when I say that I come back without one good photograph.'[43] Although Chapman was to have more success in the future, it is notable that as early as 1860 the failure to secure a photographic record was seen as tantamount to the failure of exploration itself.

The photographs from the Zambezi Expedition were, in retrospect, some of the most influential outcomes of an otherwise rather disastrous enterprise. For a start, the official role given to photography on the expedition signified an important recognition of the relevance of the technique on scientific expeditions. Indeed, Kirk's experience with photography on the Zambezi was used to furnish the first section on photography in the second edition of the RGS's *Hints*

14 Albert G. S. Hawes, 'Camp of the R.G.S. expedition in the grounds of H. M. Consulate Nyassa', 1886.

to Travellers in 1865.[44] This highly influential and long-running series evolved from the efforts of influential geographers to provide advice to explorers on equipment and observational techniques in order to improve the applicability of their labours to geographical science. The manual was part of a wider concern with training explorers to observe more fully, clearly and accurately which amounted, in effect, to an attempt to make them photographic.

Following subsequent efforts by influential figures such as Francis Galton the RGS eventually appointed John Thomson as its Official Instructor in Photography in 1886. The post officially confirmed the recognition of the medium as a legitimate form of scientific currency. Along with the fact that explorers could increasingly rely on gelatine film and hand-held cameras, this development confirmed that photography was well established as a tool of geographical exploration. As Thomson himself put it in 1885, 'No expedition, indeed, now-a-days, can be considered complete without photography to place on record the geographical and ethnological features of the journey.'[45]

Photography was certainly used at this time to document expeditions to other parts of Africa where Livingstone's vision was being put to greater effect. The 1885–6 RGS expedition to Nyasaland, led by Thomas Joseph Last, was recorded in a series of photographs by Albert Hawes, then Consul for the territories adjacent to Lake Nyasa, who accompanied Last on various expeditions in 1886.[46] A photograph of their camp in the grounds of the consulate in Nyasa (illus. 14), where Last was accommodated during the wet season, conveys the hierarchy of the expedition force, with Last seated at the centre. While the surviving photographs of the Zambezi Expedition, over two decades earlier, are notable for the invisibility of the explorers themselves, Hawes's photograph depicts the expedition as a force of order in the landscape, with the sextant displayed prominently in the foreground and the RGS flag marking out the expedition camp in the midst of tropical jungle.

Later photographs made by British colonists in British Central Africa emphasized the ways in which the landscape had been cultivated and improved for civilized pursuits. For example, an album of photographs made around Blantyre in British Central Africa in 1900–5 is dominated by scenes of roads, churches, hospitals, government buildings, coffee plantations and British outdoors recreation, from picnics to croquet.[47] These are in stark contrast to scenes of dark, tangled vegetation,[48] the presence of which is some measure of the continued hold of the image of 'darkest Africa' on British imperial imaginations.

2 Framing the View

Explorers were neither unique nor the most advanced in their use of photography to capture the distant realities of Empire. Commercial photographers keen to exploit the European demand for exotic scenes set out as early as the 1840s to 'discover' foreign lands photographically.[1] Nöel-Marie Lerebours, a French optician, commissioned a number of photographers to gather images of monuments and sites from various parts of the world, including North Africa and the Near East, and these formed the basis of engravings and acquatints published in *Excursions daguerriennes* between 1841 and 1844.[2] The increasing numbers of international travellers in the second half of the nineteenth century, particularly to locations such as Egypt and the Holy Land considered exotic by the well-to-do British tourist, provoked a rash of pictorial activity and exploration.[3] Scenes of 'the Orient' were rendered increasingly familiar to the British public through the work of painters such as David Roberts and photographers such as Francis Frith.

Frith toured Egypt, Sinai and Palestine between 1856 and 1859 documenting the monuments and landscape before returning to set up his well-known photographic firm based in Reigate.[4] Like many commercial travel photographers, Frith saw himself as both an artist and scientist; his photographic views were never intended simply as art and he was as concerned with the scientific nature of his work as he was with its picturesque portrayal of landscapes and ancient monuments. Thus in 1859, when he was planning to explore the White Nile in his own steamer, Frith approached Francis Galton, then Honorary Secretary of the RGS and well known as an explorer of south-west Africa, for advice and information. Galton had journeyed on the Nile in 1844, was familiar with Frith's work and thought him an excellent photographer.[5] Frith was one of a range of British travel photographers, including Samuel Bourne and John Thomson, whose work may be seen as part of a discourse on imperial geographical exploration. Their photographic expeditions were animated by a sense of discovering the unknown and they prided themselves on the arduous conditions they endured to secure their photographs; indeed, to many viewers this enhanced the novelty, scientific worth and artistry of the images.

Commercial photographers expended considerable energy capturing views of landscape. Some thought this was a particularly strong feature of British photography. One reviewer at the British Department of the Paris Exhibition of 1867, writing in the *British Journal of Photography,* declared landscape photography to be 'our *specialité*', since 'English men love the country. We are a nation of excursionists.'[6] Indeed, the prospects for excursions opened up by imperial expansion profoundly amplified the Victorian taste for landscape photography. Moreover, the very idea of Empire depended in part on an idea of landscape, as both controlled space and the means of representing such control, on a global scale.[7] For landscape was not something already 'out there' waiting to be recorded on glass plates or sketchpads. Rather, it amounted to a particular way of picturing and imaginatively appropriating space by a detached, individual spectator.[8]

The historical evolution of the landscape idea shows that it is especially suited to the discourse of imperialism which, as W. J. T. Mitchell has noted, 'conceives itself precisely (and simultaneously) as an expansion of landscape understood as an inevitable, progressive development in history, an expansion of "culture" and "civilization" into a "natural" space in a progress that is itself narrated as "natural" '.[9]

Moreover, as Bernard Smith's account of European visions of the South Pacific shows, the landscape idea was closely associated with European exploration and expansion overseas well before the advent of photography.[10] It is also notable that interest from the mid-eighteenth century in the possibilities of permanently securing images produced by the camera obscura – what might be thought of as a discursive desire for photography – was predominantly concerned with the appropriation of views of nature and of landscape.[11] Landscape photography as practised in the 1860s, 1870s and 1880s by British professional travel photographers was, as I go on to show, particularly suited to the naturalization of both the landscape idea and an imperial way of seeing.

Many nineteenth-century landscape photographs made by professional photographers have been considered exclusively as aesthetic objects. However, commercial photographers worked in a variety of contexts, including those of 'scientific' expeditions and surveys. For example, the photographs of the American West by Timothy O'Sullivan were made in an expeditionary context, not merely as 'landscapes' to be enjoyed as tasteful scenes, but as 'views' to supply information for the United States government on the geography and resources of the unexplored West.[12] Likewise, the history of landscape photography of the Canadian west in the second half of the nineteenth century shows how photography was closely applied to topographical surveying and the marketing of the Canadian Pacific Railway, completed in 1885.[13] In British Columbia between 1858 and 1885, for example, landscape photographers focused on the frontier wilderness, railways, mining and settlement. This period, from the Fraser River gold rush to the completion of the Canadian Pacific Railway, witnessed the dramatic extension of European colonial authority, commerce and settlement in British Columbia and photographers produced landscape scenes which captured the sense of colonial progress and British colonial identity.[14] Professional photographers were also employed on official expeditions and survey parties. In the case of British Columbia, for example, the professional photographer Frederick Dally documented the 1866 circumnavigation of Vancouver Island by HMS *Scout* and another professional photographer, Edward Dossetter, accompanied Dr Israel Powell, Superintendent of Indian Affairs, on the voyage of HMS *Rocket* up the Stikine River and to the Queen Charlotte Islands in 1881.[15]

Despite the clear evidence that commercial landscape and 'art' photographers also operated within scientific and government surveys, it would be incorrect to place a categorical division between photography's 'discursive spaces', notably between the 'view' in science and the 'landscape' in art.[16] The concept of 'landscape' included that of 'view' and operated within a number of contexts, including artistic genre and scientific record.[17] Nor is there anything startlingly new in this; many commercial photographers were following in the tracks of late-eighteenth-century picturesque artists, such as William Hodges and Thomas Daniell, who, as well as being accomplished and respected artists, explicitly related their work to colonial exploration and natural philosophical inquiry.[18] What is interesting then is the way in which British landscape photographers of the 1860s, 1870s and 1880s disguised their dependence on pictorial convention in order to promote photography as an objective record of sight, in the process reinscribing imperial landscape as a natural way of seeing.[19]

Samuel Bourne in India, 1863–70

The global reach of travel photographers was such that in 1863 a young English photographer, Samuel Bourne (1834–1912), could declare in the *British Journal of Photography (BJP)*: 'There is now scarcely a nook or corner, a glen, a valley, or mountain, much less a country, on the face of the globe which the penetrating eye of the camera has not searched.'[20]

Still, the romance of exploring distant reaches of the globe with a camera was powerful enough to lure Bourne, then an enthusiastic amateur photographer, away from his job as a bank clerk in Nottingham to embark on a seven-year career as a professional photographer in India. Photography in India, as Bourne himself recognized in 1863, was 'least of all, a new thing'.[21] In particular, the events of the Mutiny/rebellion of 1857–8 and the ensuing political changes generated a renewed demand within Britain and British India for photographic images of the subcontinent. Following in the wake of this interest, Bourne joined forces with the experienced photographer Charles Shepherd to establish a photographic firm at Simla. This hill station in the Himalayan foothills was the popular summer residence of the Indian government and itself provided a significant market for Bourne and Shepherd's views and portraits. By the time Bourne left India in November 1870, photographs from the firm were sold through outlets in most of India's major cities, as well as in London and Paris. Ample sales allowed the firm to open further studios in Calcutta (1867) and Bombay (1870).

Much of their success depended upon Bourne's reputation as a photographic artist. Although his work covered a number of fields of interest, including architectural and portrait photography, he was particularly well known for his landscape views and for the series of photographic expeditions he undertook in the western Himalayas in 1863, 1864 and 1866. His frequently arduous photograph-gathering expeditions, which he described graphically in a series of articles for the *BJP*,[22] were crucial in establishing his contemporary standing. More recently they have helped make his reputation as one of the great pioneers in the history of photography.[23] However, to consider Bourne solely in terms of his contribution to photographic artistry has given rise to a limited understanding of the contexts in which his photographs were made and displayed. In particular, the significance of these expeditions, and his photographic practice generally, as exercises in colonial representation has often been overlooked.[24]

Bourne's photographs were greatly praised in India and Britain for their technical and compositional qualities. In 1869 the *BJP* agreed with one reader's enthusiastic assessment of them as 'exceptionally excellent'.[25] As one reviewer of Bourne's pictures at an exhibition that year in India declared, 'These are pictures not to be doubled up in a coarse scrap-book, but framed for the adornment of the drawing-room.'[26] To be sure, Bourne intended his photographs to be appreciated as objects of beauty and aesthetic contemplation. Yet his landscape views of northern India were lauded in their time not only for their pictorial eloquence and technical quality but also because they had been made on particularly arduous expeditions in a relatively unknown portion of the British Empire.[27] In common with other Victorian commercial photographers, Bourne infused his photographic travels with the spirit of geographical adventure and discovery. His photographs were consequently collected and admired by those interested in acquiring specific visual knowledge of the geography of northern India as well as those wishing simply for pictures of India with which to adorn their drawing-room. Thus in addition to being contemplated in private, Bourne's photographs were displayed publicly at exhibitions of photographic art, at learned and scientific societies and at international exhibitions, where they were viewed alongside those made by explorers such as James Chapman.

15 Samuel Bourne, 'View on Dal Canal, Srinagar, Kashmir', 1866.

Samuel Bourne had very definite ideas about what constituted an aesthetically beautiful, as well as a technically good, photograph. His 'View on Dal Canal, Srinagar, Kashmir' (illus. 15) contains many of the ingredients he regarded as essential in constructing a picturesque image: an expanse of water, with many reflections; wooded banks in the background and foliage in the foreground, allowing for a play of light and shade, and a carefully posed figure to draw the viewer's gaze into the image. The aesthetic conventions of the picturesque which Bourne followed had been formulated in the late eighteenth century by critics such as William Gilpin, whose guidebooks described how representations of nature should be organized. Such rules of picture making were also followed by landscape painters and engravers such as Thomas and William Daniell, who worked in India between 1786 and 1793 and who made extensive use of the camera obscura.[28] Thus by the 1860s the travel photographer in India could follow a well-trodden path made by Anglo-Indian tourists and artists in search of the picturesque. Through photographic travels such as his trip to Kashmir in 1864, Samuel Bourne was thus adding a photographic gloss to an established picturesque iconography of India.[29] Yet the authority conferred on photography to capture truthfully scenes of nature gave it a power greater than that of engravings or paintings to confirm and naturalize the landscape aesthetic.

Nevertheless, as Bourne himself recognized, the camera was certainly not new to the British in India. In 1859 John Murray had produced a series of views of the North-Western Provinces.[30] Two years later Captain Melville Clarke had made a series of photographs on his journey from Simla to Ladakh and Kashmir.[31] Although his photographic 'views', published with brief accompanying notes in 1862, are not as technically sophisticated as Bourne's, they focus on strikingly similar objects. Clarke's photographs of Simla, 'the favourite Sanatorium of the Himalayas', as well as the topographic features surmounted on his journey and picturesque scenes in Kashmir, are all echoed in the photographs Bourne made three years later.[32] Both Bourne and Clarke were drawn to the conventional attractions of Srinagar, the capital of Kashmir, such as the Shalimar Gardens and various picturesque bridges which reminded them of a bygone London.[33] Clarke used photographs to testify to the architectural achievements of the Mogul Empire, which he contrasted with 'the rude and perishable structure of the present age in Cashmere'.[34] Similarly, Bourne could not help regretting that 'such a noble but misgoverned country should not have remained in the hands of the English when it was once in their possession'.[35] Indeed, Bourne's view of the appropriateness of English rule in Kashmir was partly measured through Kashmir's suitability for photography.

Bourne had initially been disappointed by the picturesque potential of India. Upon first arriving in Simla in 1863 he complained that as well as there being 'no lakes, no rivers, and scarcely anything like a stream in this locality, neither is there a single object of architectural interest, no rustic bridges, and no ivy-clad ruins'.[36] He was thus delighted to find many of these picturesque features during his visit to Kashmir in 1866. In addition, he found a climate largely free from wind, cold or dust - elements which so often bedevilled his photography. It was no coincidence that having found Kashmir perfect for photography Bourne was reminded 'very forcibly of the hills and valleys, green fields, parks and pastures of England'.[37] Indeed, to discover beautiful landscape in India was, for Bourne, commensurate with finding the ideal English view in a different environment. 'Indian landscapes,' he wrote in 1864, 'I do not think will ever compare with English; not because the photography can not be as good, but because the scenery is not so beautiful or so well adapted for the camera.'[38] In Bourne's opinion, 'no scenery in the world is better or so well adapted for photography, on the whole, as that of Great Britain'. Speaking of the photographer's ideal he wrote: 'If he could only transport English scenery under these exquisite skies, what pictures would he not produce!'[39]

In effect, however, like many Victorian travel photographers, Bourne was embarking on just this kind of transformative project. By imposing the aesthetic contours of 'English scenery' on to foreign environments he was familiarizing and domesticating a potentially hostile landscape. Indeed, the reality that he claimed to be revealing truthfully was in fact one of his own culture's making.

For Bourne the practice of photography was a highly disciplinary pursuit.[40] His photographic explorations in the Himalayas and Kashmir were organized and executed with the ardour of an intrepid explorer, or military commander. Not only did he struggle to operate his photographic apparatus in a harsh environment and climate, but he depended on numerous porters to carry his equipment and supplies. Bourne set off on his nine-month expedition to Kashmir in 1866 with a retinue of more than fifty people: forty-two 'coolies' (porters), a staff of six servants and six 'dandy bearers'. Referring to the latter, he was anxious to point out to his readers that he was no dandy but that it was the name of the vehicle he was carried in when he got tired.[41] Bourne's expedition was certainly not lightweight. His photographic equipment alone made up twenty full loads, catering for every eventuality for, as the *BJP* noted in praise of Bourne's achievements, 'in the wilds of Cashmere, or far up among the Himalayan mountains, there is no well stocked chemist's shop to apply to when the dark hour arrives'.[42]

The wet-plate collodion process Bourne used entailed coating, sensitizing, exposing and developing a glass plate in immediate succession. This required the transport and repeated erection of Bourne's pyramidal dark-tent which, with a height and width each of ten feet, was no simple task. In addition Bourne insisted that his expeditionary equipment included tents, bedding, sporting requisites, books, camp furniture and a good supply of Hennessy's brandy. It was advisable to take as many luxuries as possible, he explained, 'seeing that I was sometimes for two months in some solitary and remote district without ever seeing a European, talking nothing and listening to nothing the whole time but barbarous Hindostani'.[43]

Although undertaking such an expedition inevitably required considerable organization and equipment, it was Bourne's cultural baggage that weighed more heavily on the order of the expedition than anything else. His often blatant contempt for the people he encountered and upon whose labour and resources he depended was reflected in the ease with which he wielded any nearby stick when members of what he referred to as his 'little army' shirked their duty or displayed signs of 'mutiny'.[44] Moreover, the porters and guides upon whom he depended feature in his photographs primarily as picturesque figures for graphic charm and scale rather than as individuals in their own right.

More generally, the inclusion of local inhabitants in Bourne's pictures

16 Samuel Bourne, 'Kashmir Women', 1866.

depended upon their total acquiescence to his aesthetic. He complained, for example, that the idyllic conditions for photography during his earlier visit to Kashmir in 1864 – considerably helped by the servants and boat he had at his disposal – was marred only by the

obstinacy of the natives when I wanted to introduce them into my pictures. By no amount of talking and acting could I get them to stand or sit in an easy, natural attitude. Their idea of giving life to a picture was to stand bolt upright, with their arms down as stiff as pokers, their chin turned up as if they were standing to have their throats cut.[45]

Although Bourne was interested primarily in a landscape aesthetic that minimized humans, he was also drawn to the indigenous Indian population, particularly if they held some picturesque attraction and could be coerced into sitting for him. While in Kashmir, for instance, he showed a particular interest in and admiration for the 'wives and daughters of the Hindoos', who, he said, were 'fairer than the Mohammedan women'. Bourne told his readers, 'It is only when you come upon them unawares that you can see them properly.'[46] Consequently, he depended on the help of the English Commissioner resident at

Srinagar to give orders for 'the best looking nautch girls' to be collected in order to be photographed (illus. 16).[47] Despite – or perhaps because of – such coercive tactics, Bourne produced a rather unimaginative composition with little evidence of an 'easy natural attitude' among the women. Nevertheless, Bourne's attraction to them as photographic subjects and his use of colonial authority to arrange them before his lens are symptomatic of a wider currency in sexualized imagery of the colonial exotic. 'Nautch girls' were long regarded by the British as equivalent to high-class prostitutes and were treated with a mixture of fascination and disdain. They were a popular subject for photographers, who made pictures of them posed in their traditional costumes, occasionally in mid-dance, further amplifying the stereotype of the sexually exotic Oriental woman. Such images found their greatest expression in the salacious and pornographic photographs of the colonial harem manufactured by European commercial photographers, particularly in North Africa and the Middle East.[48] Drawing on an established iconography of the Orient in painting, literature and architecture,[49] photographers invented their own 'Orient' in the studio with props, backdrops and the directed poses of scantily dressed 'Oriental' female models. This photographic imagery exploited existing associations between the Orient and sex, making apparently realistic evidence available on a ever larger scale. Titillating photographs with titles such as 'Favourites of the Harem' were thus marketed by the international photographic firm and news agency Underwood and Underwood well into the twentieth century.[50]

Although many such photographs were made in studio settings, they were by no means unrelated to landscape views. Indeed, scenes of tropical vegetation – improvised with paint and plants in the studio as well as in outdoor locations – were employed frequently as backdrops for images of exotic women. Among the portraits of people from Ceylon made by the commercial photographers W. H. L. Skeen & Co. in the 1870s and 1880s are several which picture women against a background of luxuriant tropical vegetation.[51]

While Samuel Bourne did not eroticize so overtly the landscapes of northern India, he did project them as fleeting beauties to be conquered by the male and white power of endurance. His most energetic assault on the Himalayan topography came in his third and final expedition in 1866, when he set out, in his own words, to 'explore the rich valley of the Beas River through Kulu, penetrate into the wild and desolate regions of Spiti as far as the borders of Thibet, thence, via Chini and the Buspa Valley, to the source of the Ganges'.[52] Although he claimed his object was 'purely pictorial' with, he assured his readers, 'no pretensions to *scientific* travels',[53] these geographical feats and the language in which they were described were the very stuff of heroic exploration. Far from being incidental to a wandering search for the picturesque, they were exercises in imperial assertion which directly correlated with Bourne's aesthetic preoccupations.

From the start, Bourne's expeditions were intended to expose unknown geography to scrutiny. Describing his initial ten-week expedition of July–October 1863, he explained how he wanted to explore the 'interior' of the Himalayas, to 'see what elements of beauty and grandeur lay concealed'.[54] His photographic representation of the landscapes and inhabitants of the western Himalayas should therefore be seen as part of a related project of visual surveillance; a means of surveying spaces and peoples from an imperial point of view. This is shown, for example, in the use of photography by Philip Henry Egerton, whose illustrated *Journal of a Tour through Spiti, to the Frontier of Chinese Thibet* (1864) began:

The Himalayan Mountains, which in former years have attracted the Tourist, the Geologist, the Botanist, and the Sportsman, are daily acquiring a more extended interest as containing considerable quantities of forest, or jungle land, which if cleared and brought under tea cultivation, are capable of affording to many of our countrymen, if not vast fortunes, at least comfortable affluence, with cheerful and pleasant occupation in a good climate.[55]

In 1863 Egerton, Deputy-Commissioner of Kangra, had explored a route through the Province of Spiti (part of the British-ruled Kangra District) to the frontier of Chinese Tibet in an attempt to enlarge the share of commercial traffic undertaken by Britain's Indian territories. In particular, he envisaged establishing a new route to link the latter directly with Yarkund and Chinese Tibet, so thwarting the Maharaja of Kashmir's monopoly of the wool and shawl trade. Photographs of 'people and places probably never before delineated with accuracy' were a vital part of Egerton's explorations, which he hoped would 'bring manufactures into the heart of Central Asia, extending civilization to the barbarous hordes which people those vast tracts, and enriching the manufacturers, exporters and carriers of European produce'.[56] Egerton's use of photography was thus finely tuned to his official imperial mission.[57]

Like Bourne, Egerton travelled in some 'state and splendour'. Indeed, his private equipment, including his 'Swiss-cottage tent' and photographic apparatus, required thirty-six men to carry it. In addition he took numerous articles to give to local dignitaries as presents. He was accompanied by servants, interpreters and guides, as well as a sergeant of police and four constables. Together with his travelling companion Mr Heyde, a Moravian missionary and Tibetan scholar, and Heyde's servants, Egerton's party numbered over seventy.[58] Despite their large expeditionary parties, both Egerton and Bourne saw themselves as lone pioneers entering unknown and uncivilized territory. Thus when Bourne ran out of photographic chemicals after his first expedition, he headed home, 'anxious to get back to civilization and English society'.[59]

To be sure, Samuel Bourne was not participating in an official expedition like Egerton. However it would be inaccurate to assume that Bourne's artistic motivation somehow absolved his work of imperial ideology or influence. This implies that the political context of imperialism merely compromised an otherwise pure cultural or aesthetic realm, when it was in fact a major source of imaginative power. While the colonial ambitions of Bourne's expeditions were perhaps less overt than Egerton's, they were just as present in his photographic practice.

Bourne's expeditions were reliant upon wider networks of knowledge and systems of power, without which he would have been unable either to have secured porters and supplies or even to have planned and executed his route. Moreover, on his 1866 Himalayan expedition Bourne followed the very same route that Egerton had set out on some years earlier, relying on the same Great Trigonometric Survey map that had been prepared for Egerton's *Journal* of 1864.[60] Bourne followed Egerton's route along the valley of the Beas River to Sultanpoor, the capital of the district of Kulu, which had been under British control since 1846. Like Egerton before him, he avoided the Rotang Pass and instead entered the district of Spiti via the Hamta Pass, which he photographed. Egerton had also photographed this pass (illus. 17), a print of which he included in his *Journal*.[61] The photograph was intended as a geographical record, part of a journey whose route was plotted and measured precisely. Although Egerton was also concerned with capturing views of the surrounding scenery, poor weather in July and on his return in September stopped him making photographs from the top of the pass.

Bourne, on the other hand, could afford to be more patient when he visited the area. Although cloud obscured the mountains when he initially crossed the pass, he camped below it and returned on the following two days, during which he made five photographs including one he titled 'The Hamta Pass between Kulu and Spiti, elevation 14,300 ft.' (illus. 18).[62] Although Bourne was more concerned than Egerton with depicting generalized picturesque scenes, his detailed title locates the photograph as a geographical document even more precisely than Egerton's. Looking only at the photographs, there seems little to distinguish them. Indeed, Bourne was quite literally following a path beaten by Egerton. By the time Egerton returned from Spiti, taking the same route, in September 1863, he had given orders for the improvement of the pass and found 'a good path made along the side of the ravine'.[63] This was the very path Bourne relied upon three years later.

Following his crossing of the Hamta Pass, Bourne, again like Egerton before him,[64] explored the Shigri Glacier, making photographs and speculating on its geological implications. Bourne was accompanied during this part of his expedition by Dr George Rankin Playfair of Calcutta, brother of the well-known naturalist Dr Lyon Playfair. George Playfair was a keen botanist and geologist, and his knowledge added much to the interest of the expedition for Bourne. It

17 Philip Henry Egerton, 'The Hamta Pass', *Journal of a Tour Through Spiti* (1864).

18 Samuel Bourne, 'The Hamta Pass between Kulu and Spiti, elevation 14,300 ft.', 1866.

its dramatic nature and partly because it was said to express 'a national characteristic of the Chinese race', namely their apparent hatred of foreigners and the outside world in general.[111]

Thomson's interest in photographing the differences and boundaries between East and West was by no means limited to his travels in China. In the autumn of 1878 he took advantage of a timely opportunity to explore the island of Cyprus, which had been occupied by Britain only a few months earlier. In his resulting two-volume work, *Through Cyprus with the Camera* (1879),[112] Thomson claimed to have produced 'incontestable evidence of the present condition of Cyprus' on the eve of British colonial rule. His photographs, together with the written observations which accompanied each one, offered 'faithful reproductions' of the 'topography' and 'resources' of a country 'woefully wrecked by Turkish maladministration'. Despite its present poor condition, Thomson argued, Cyprus was far from exhausted and his photographs supplied evidence to predict great colonial prospects when, 'under the influence of British rule, the place has risen from its ruins'.[113]

It was as a forecast of future colonial progress that Thomson made landscape photographs of Cyprus. 'The Sea Shore, Larnaca' (illus. 24) is a fine example of this colonial, improving vision. With a solitary figure for scale, the photograph presents an elevated view along the shoreline towards Larnaca. Thomson referred to the nearby salt ponds as 'nothing more than malarious swamps', which ought to be drained to improve health and to reclaim land for the future expansion of the city.[114] Along with the draining of marshes, the restoration of forests[115] and the rebuilding of harbours,[116] Thomson sought to restore the ancient Greek architecture and the Gothic churches which he believed to have been ruined by the Turks.

His photograph of 'The Front of St Katherine's Church (Now a Mosque), Famagosta' (illus. 25) is emblematic of many of his views of urban landscapes in Cyprus, in which he contrasts a 'noble specimen of Gothic architecture' with the Turkish developments which have 'sadly mutilated' it:

> The tower on the left is Turkish, and tells us at once that the splendid old cathedral pile has been turned into a Mosque; a motley range of modern hovels have also grown up under the shadow of the church. On the left for example, stands one of the most imposing specimens of the present architecture of the place. It is a café, propped upon an old Gothic arch and adorned with a flag staff. Here worshippers at the shrine of the Prophet meet and sit for hours, smoking their hookahs, and drinking their coffee in silence; for they have long ago exhausted all the subjects of conversation that so lonely a spot can supply.[117]

As his comments on Turkish habits suggest, Thomson considered Famagusta to have fallen into decay because of the Turks and envisaged, under British rule, 'the erection of a new city'.[118] He presented the same picture of ancient glories hidden by recent decay in his representation of the cathedral in Famagusta, again converted into a mosque, where 'Mussulmans worship above tombs where Knights, Crusaders, and Venetian nobles rested'.[119] In other parts of Cyprus, for example in Nicosia, Thomson similarly contrasted explicitly the old Gothic style with what he disparagingly termed 'Turk mud-ine'.[120] His improving eye anticipated the clearing of refuse, the repair of roads, the sanitization of city spaces and the cleansing of 'all the odours of decay peculiar to Eastern towns'.

What was at stake here, for Thomson, was the difference between the despotism and misrule of the Oriental Turks on the one hand and the improving capabilities of European Cypriots under British colonial supervision on the other. This opposition between Oriental and Occidental was visualized in photographs and accompanying text which asserted the right and duty of the British to rule Cyprus. Like much commercial landscape photography, Thomson's work was not only driven by a broad concern with recording the improving strides of British ideas, culture and civilization but was also part of a colonizing movement, establishing visual guidelines and justifications for more palpable forms of colonial control. That said, the extent to which Thomson's faith in and advocacy of British colonial rule conditioned the interpretation of his photographs depended upon their close reading in conjunction with his written text. Moreover, this was a reading which could not be guaranteed, particularly since such photographs often had existence beyond any accompanying text or even caption. The exhibition of Thomson's photographs in spaces of science and of art thus opened them up to a range of alternative interpretations, as simply scientific records or as exercises in pictorial beauty, in which imperial preoccupations were not necessarily primary.

Picturesque Travel and Imperial Landscape

The work of Samuel Bourne and John Thomson is by no means unrepresentative of mid-Victorian commercial travel photographers working around the world. Similar processes can be seen in the work of the Australian commercial photographer John William Lindt (1845–1926). With his studio based in Melbourne from 1876 to 1894, Lindt acted as official photographer to Sir Peter Scratchley's New Guinea Expedition of 1885 shortly after a British Protectorate had been proclaimed over South-East New Guinea. Inspired in part by Thomson's *Illustrations of China and Its People,* Lindt determined to produce a 'book of travel, entirely illustrated by artistic views and portraits taken direct from nature', resulting in his *Picturesque New Guinea* (1887) in which his photographs were reproduced in the new reproduction process of autotype.[121] Like Thomson, whom he had met in London, Lindt was a Fellow of the RGS and put forward his work as a contribution to geographical exploration, arguing that 'artistic photography' was 'the legitimate and proper means to

26 John William Lindt, 'Near the Camp, Laloki River', *Picturesque New Guinea* (1887).

show friends at home what these foreign lands and their inhabitants really look like'.[122] His photograph 'Near the Camp, Laloki River' (illus. 26) is one of a number taken during this expedition which focus on groups of indigenous people arranged in a natural landscape setting. Other photographs record the camps, members of Sir Peter Scratchley's party and views of tropical vegetation, animals and landscape. In addition, Lindt took an anthropological interest in the houses, social rituals and appearance of the indigenous inhabitants. Indeed, he was well versed in composing scenes which combined the ethnographic with the picturesque, having already made something of a name for himself with his studio portraits in the 1870s of Aborigines posed against painted landscape backdrops. Similarly, he carefully composed the view 'Near the Camp, Laloki River' to emphasize the picturesque elements of both the landscape and the inhabitants. In his text Lindt described the peoples around Laloki River as being 'an indolent and filthy race', though he later commented favourably on their industry and ingenuity.[123] He was not averse to exploiting the aesthetic appeal of the indigenous inhabitants, who were included or omitted from his photographs depending upon his requirements for making picturesque views. Indeed, another photograph shows the same scene from slightly further away emptied of all human figures.[124] Both scenes were captured for sale to a domestic audience in Melbourne and a wider colonial world beyond. Rather than etching his name or a number on the negatives, as Bourne and Thomson both usually did, Lindt instead placed a small wooden sign with 'LINDT, MELBOURNE, COPYRIGHT' in white lettering on the ground in the field of vision, staking his claim to the landscape and marking permanently the scene as his property and copyright.

The *Picturesque New Guinea* project as a whole – which Lindt dedicated to Queen Victoria – amounted to a classic piece of colonial advertising.[125] Indeed, like Thomson, he put forward his photographs as evidence of his written confidence in the benefits of colonial rule. His photographs of a mission house, for example, gave visual evidence of how 'even in savage New Guinea the blessed light of the Word of God is gradually dispelling the darkness of barbarism and cannibalism'.[126] To Lindt, photography was an essential means of opening up such landscapes and their inhabitants to colonial improvement and the order of civilization.

Lindt's work, like that of Bourne and Thomson, shows how landscape photography, like topographial surveying and cartography, was frequently represented as an external view and instrument of visual colonization. As well as shaping images of indigenous people, photography tended to naturalize the landscape aesthetic as a rational, distanced way of viewing and organizing space. In the terms exercised by their makers, landscape photographs such as those made by Bourne, Thomson and Lindt translated unknown spaces into familiar scenes, opening up distant territory to imperial eyes. Yet ultimately their effect as imperial signs depended too upon their circulation and reception across a range of spaces, from the framed picture on a drawing-room wall or in an art exhibition to a set of volumes in a scientific society or government department.

Photography was of course only one medium used in the picturing of foreign landscapes. Indeed, as I have noted, it was often used in conjunction with other forms and conventions of representation, from travel writing to cartography. However, perceived as an 'Art–Science', photography incorporated a role as 'handmaid of the sciences' with an aestheticizing project of global proportions. As one photographic enthusiast put it in 1860:

> Our indefatigable countrymen are ascending the Nile, tracing the course of the Zambezi, navigating the Ganges, the Yang-tse-Kiang, the Mississippi, climbing the Alps, the Andes, the Himalayas, in fact wandering to every region of the habitable globe in search of the beautiful and the picturesque.[127]

Photography was particularly suited to such an enterprise of global survey, since, as Susan Sontag has noted, 'from its start, photography implied the capture of the largest number of subjects. Painting never had so imperial a scope.'[128] Indeed the camera, the 'pencil of the sun', seemed destined to reveal all the mysteries of the world to an expanding imperial gaze.

3 The Art of Campaigning

From the 1850s an increasing number of calls were made, especially in military and photographic journals, for the application of photography to military purposes. The role of photography in military operations was envisaged as a logical extension of the general scientific function of this 'handmaid of the sciences'.[1] Thus in 1854 the *Art Journal* had reported on the possibility of collaboration between photography and military engineering operations in the recently declared war in the Crimea.[2] Later in the same year the British Association for the Advancement of Science heard details of specialized camera equipment designed for military application by the army and navy, including an 'army field camera' mounted on a large pair of wheels so that it might be easily manoeuvred by a single operator.[3]

Enthusiasm for the military applications of photography stemmed in part from the increasingly prominent place of warfare in the Victorian imagination, particularly where Empire was concerned. Indeed, the Victorian Empire was sustained in large measure by regular shows of military force. With the sole exception of the Crimean War the many military campaigns during Queen Victoria's reign were fought entirely against non-European enemies. One veteran of campaigns in Afghanistan and India, Major Charles Callwell, later termed these conflicts 'small wars', describing them as 'a heritage of extended Empire' which 'dog the footsteps of the pioneer of civilization in regions far off'. Callwell applied his term particularly to 'expeditions against savages and semi-civilized races by disciplined soldiers'.[4] The heroes and villains of such 'small wars' certainly provided exciting subjectmatter for all manner of Victorian cultural forms, from popular literature and art to the spectacular re-enactments of battles at events such as the Royal Naval and Military Tournament.[5] It was in such displays that the military, and a militarist spirit, came to occupy an unprecedented place in Victorian culture.[6]

One of the most spectacular Victorian 'small wars' in which photography played a significant role was the Abyssinia Campaign of 1867–8. This highly celebrated campaign was launched ostensibly to rescue a handful of Europeans, including the British Consul, imprisoned by the Emperor Theodore of Abyssinia (Tewodros II) following a series of diplomatic misunderstandings.[7] Hence Major Callwell placed it in his third class of 'small war': a campaign to 'wipe out an insult, to avenge a wrong, or to overthrow a dangerous enemy'.[8] Conducted by a military force of nearly 13,000 men from Britain and India, the campaign involved an expedition of some 400 miles from the coast at Annesley Bay to Theodore's highland stronghold of Magdala, whereupon the hostages were rescued and the city destroyed. Before the British forces could reach him, Emperor Theodore shot himself with a pistol Queen Victoria had given him as a gift some years before, much to the interest of the troops and assorted press accompanying the campaign.

This dramatic rescue mission received unprecedented publicity in Britain. With extensive provision made for reporters from major newspapers, including G. A. Henty (*Standard* and *Pioneer*), Viscount Adare (*Daily Telegraph*), C. Simpson (*Illustrated London News*) and H. M. Stanley (*New York Herald*), there was little chance that the campaign would go unreported. Moreover, both Henty and Stanley went on to produce book-length accounts of the campaign.[9] As a massive media 'performance' – indeed the *Times* correspondent, Dr C. Austin, referred to it as such[10] – the campaign demonstrates the increasing attention given by the press and British public to the role of the army abroad. With such media attention, detailed military planning and the considerable expense of the operation – later calculated to have cost some £9 million – this campaign proved an effective means by which Disraeli could focus the British public's attention on a distant imperial adventure, providing a popular distraction from a range of domestic crises.[11]

These considerable investments were undoubtedly important motivating forces behind the official recording of the campaign in photographs. Two bulky sets of photographic stores and equipment (of which only one was actually used) were sent from England at the suggestion of the director of the Royal Engineers' Establishment at Chatham. The equipment was supervised in the field by a chief photographer, Sergeant John Harrold, and seven assistants. Besides their other duties, the Royal Engineers used the camera to record scenes of the expeditionary forces, portraits of officers and landscape views. Although it is not known how many such photographs were made in total, a series of seventy-eight, including landscape views, camp scenes, sketches and portraits were subsequently assembled into albums and presented to various worthy institutions of government and science, from the RGS to the Foreign Office, by the Secretary of State for War in 1869.[12] A number of the photographs were also used, along with drawings by various officers, as a basis for the illustrations in the official *Record of the Expedition to Abyssinia*.[13]

In this context the photographs made by the Royal Engineers in Abyssinia make up an official record of the campaign as seen from the perspective of a great imperial power overcoming harsh geographical conditions to vanquish a tyrannical ruler. Carefully posed studies, which might just have easily been

27 Royal Engineers, 'Sir R. Napier and Staff', 1868.

made in studios in London or Bombay, portray the main characters in the drama: from General Napier and his commanding officers scanning maps (illus. 27) to groups of rescued Europeans conspicuously wearing their prison chains.[14]

As I go on to argue, the picturing of warfare in this campaign provides a useful means of considering the broader place of photography in British imperial militarist discourse. Before examining the precise roles assigned to photography within the Abyssinia Campaign, however, it is important to consider the relationships between photography and the military in more general terms.

Imperial militarist discourse manifested itself in photography in a range of ways, one of which was the rhetoric and practice of commercial photography. Soon after his arrival in India in 1863, the commercial photographer Samuel Bourne described the power of the camera in starkly militarist terms:

> From the earliest days of the calotype, the curious tripod, with its mysterious chamber and mouth of brass, taught the natives of this country that their conquerors were the inventors of other instruments besides the formidable guns of their artillery, which, though as suspicious perhaps in appearance, attained their object with less noise and smoke.[15]

Bourne's analogy between the camera and the gun is based not only on their similar appearance but on the fact that they are both weapons within a praxis of colonial power. Indeed, as I have already noted, Bourne's photographic expeditions in the Himalayas of northern India were conducted in the overweening style of the small-scale military campaign.

Commercial photographers also recorded the activities of the British army in various corners of the Empire. Indeed the firm Bourne and Shepherd sent one of their photographers, Benjamin Simpson, to record scenes from the second Anglo-Afghan War in 1878–9. Bourne and Shepherd duly presented an album of landscape views in and around the Koorum Valley to Lady Roberts, whose husband Lord Roberts commanded the British forces. The war[16] was more extensively photographed by another professional photographer, John Burke, who had been invited by the government of India to act as official 'photographic artist' and duly attached himself to the Peshawar Valley Field Force invading Afghanistan. Despite a later dispute over the terms of employment, Burke produced a fine body of photographs depicting dramatic landscapes, British troops and senior military commanders.[17]

In a number of respects the photographs of the Second Anglo-Afghan War and the Abyssinia Campaign a decade earlier follow a set of already established conventions for depicting warfare. The orderly camp scenes, formal portraits of officers and generals and landscape views are thus reminiscent of some of the earliest war photographs made by the professional photographer Roger Fenton in the Crimea (1854–5).[18] Sponsored by the Manchester publisher William Agnew and with the official patronage of Queen Victoria and Prince Albert, Fenton's record of the war presented a sanitized vision of a hopelessly managed campaign in which most of the British casualties resulted from exposure, disease and malnutrition. Fenton's photographs have little of the stark realism of William Howard Russell's commentaries in *The Times* which did so much to ignite the controversy that forced Lord Aberdeen's government to resign in February 1855. Thus, despite the claims of accuracy made for photography, in this case it hardly showed warfare more realistically than other media and indeed was a powerful means of displaying the official 'truth' of a campaign.

While the practical requirements of photography, with its long exposure times and bulky apparatus, confined it to recording the aftermath of conflict, they did not necessarily inhibit a photographer's ability to portray the violence of warfare. Indeed the more sensational aspects of warfare were popular targets for commercial photographers. Professional photographers such as Felice Beato hurried to India in 1858 to record scenes of 'the Mutiny'. His photographs of the remains – human and architectural – of the Secundra Bagh at Lucknow, several months after it had been attacked by British troops, catered for the contemporary fascination with 'the Mutiny' within Victorian art and literature.[19]

The supposed accuracy and immediacy of photography thus made it especially effective as a witness of a successful campaign. As well as having profound repercussions on military organization, the Crimean War had demonstrated the powerful influence of visual and printed media on domestic attitudes to overseas conflict. In the Abyssinia Campaign both the army and

Disraeli's government had much at stake, so arranging for favourable representation in various media was a matter of crucial importance.

Besides the activities of commercial operators, the British military itself played an important role in the spread of photography throughout the Empire. It was often officers and military surgeons who were the first to take up photography after its invention in 1839. One notable example, Dr John McCosh (1805–85), started in the early 1850s using Fox Talbot's recently developed calotype process while serving as a doctor with the 31st Bengal Native Infantry. During his military service in India McCosh took a substantial number of photographs of a range of subjects, many of which were made on particular campaigns. While serving with the East India Company's forces in the Second Sikh War (1848–9), for example, McCosh made portraits of officers and commanders of the 2nd Brigade. He also used his camera during the Second Burmese War (1852), when he served with the 5th Battery, Bengal Artillery, making a number of photographs, including some of Rangoon's pagodas, palaces and captured artillery.[20]

Although McCosh described photography as a hobby he thought it important enough in his *Advice to Officers in India* (1856) to recommend:

> every assistant-surgeon to make himself a master of photography in all its branches . . . During the course of his service in India, he may make such a faithful collection of representations of man and animals, of architecture and landscape, that would be a welcome contribution to any museum.[21]

Such sentiments represented in rudimentary form the emergence of an attitude in which India was itself regarded as a vast 'living museum'.[22] Indeed, McCosh's wide interests in topography, ethnography and travel, and his application of photography to the picturing of fellow officers, ethnological 'types', architecture and landscape should not be divorced from his duties within the British army, an organized military institution whose function was to protect the interests of a powerful imperial state. The intimate associations between scholarly observation, classification and description on the one hand and modern European imperialism – particularly through military campaigns – on the other can be traced at least as far back as the French invasion of Egypt in the late eighteenth century. Although McCosh's work is more eclectic and less well crafted than that of subsequent soldier photographers, such as Captain T. Biggs, Major R. Gill and Captain L. Tripe, his photographs are closely allied to such a tradition of Orientalist representation. This can be seen in both his earlier lithographic work in his *Topography of Assam*[23] and his later front and profile photographs of 'types', labelled in terms of sex, race and/or nationality. McCosh's concerns with geographical and ethnological classification and description thus have numerous precedents, evident not least within his own earlier work on the mountain tribes in north-east Bengal.[24]

That the work of such soldier photographers was framed by wider imperial concerns is also evident in McCosh's exhibition of photographs at the RGS in 1860, where he argued for the British colonization of territory between north-east Bengal and China.[25] That McCosh found the RGS a suitable venue in which to exhibit his photographs and to promote his schemes for British colonialism in Asia is one indication of the connections being forged between the military, colonial science and the practice of photography. It was precisely because of individuals such as McCosh that the camera was assuming an increasingly official place among the British military and in the art of campaigning.

Photography and the Military: The Royal Engineers

Within military circles, it was the Corps of Royal Engineers which provided the most receptive military environment for photography. Captain William de Wiveleslie Abney, RE (1843–1921), for example, wrote and lectured extensively on photography, earning considerable praise within and beyond the military for his researches and inventions.[26] Largely through the efforts of men such as Abney, the Royal Engineers quickly adapted photography to various tasks in their civil and military duties.[27] Photography was one of a number of practical technologies, from electricity to telegraphy, put to the military service of Britain and her Empire by these 'scientific soldiers'.[28] In the 1850s photographic equipment was dispatched to units of the Royal Engineers serving in India. Engineers of the East India Company were also encouraged to keep a photographic record of the progress of public works projects. Photography was included in the curriculum of cadets at the East India Company's Military Seminary at Addiscombe from 1855. Furthermore, a number of Royal Engineers received instruction in photography from Charles Thurston Thomson of the South Kensington Museum in the early 1850s. Following the establishment of a School of Photography and Chemistry in 1856, instruction in photography was included in the training courses at the Royal Engineers' Establishment at Chatham, which from 1850 had provided instruction and training of recruits for the whole Corps, for home as well as foreign service.[29] With the amalgamation of the East India Company's Engineers and the Royal Engineers in 1862 the photographic interest achieved a more organized and official military blessing.

By 1860 the Royal Engineers were wielding cameras on a truly global scale. Thus when Captain John Donnelly lectured at the United Services Institute in London in that year on the application of photography to military purposes, he exhibited photographs taken by parties of sappers working in Asia Minor, Panama, India, Singapore, China and Russia.[30] With such visual evidence to support him, Donnelly argued that 'photographs of a country gave a most truthful and accurate idea of it. They would do more to give an accurate idea of any particular position than yards of description on foolscap.'[31] So as well as being useful in the reproduction of maps and plans and in the illustration of the progress of military operations, Donnelly claimed that photography provided 'a perfect idea of a place'. This meant that it provided strategic knowledge of the geography of a country which might be a potential field of war.

In 1863 John Spiller, then Assistant Chemist to the War Department and Photographer to the Royal Military Repository, Woolwich, reported in the *BJP* that artillery officers' 'field days' with a photographic van on Woolwich Common gave 'proof of the value which attaches to photography as a ready means of recording the geography and military features of a country'.[32] The recognition in military circles that photography was a useful means of recording geographical knowledge provided a spur for its adoption on campaigns such as that in Abyssinia, to which I now return.

Photography in the Field

The use of photography by the Royal Engineers in the field received much attention in the pages of photographic and scientific journals during and after the Abyssinia Campaign. Such journal reports added to a wealth of more general publicity in newspapers, where the expedition was commonly praised as an application of science to modern warfare. In 1870, for example, the author of an article in *Nature* claimed that, second only to the French, the British War Department was most advanced in its application of the 'Art-Science' of photography to military purposes. This, the author went on, was one reason why the recent Abyssinia Campaign was 'one of the most wonderful feats of engineering accomplished in modern times'.[33] According to another commentator in 1868, the labours of the Royal Engineer photographers in Abyssinia formed 'no unimportant part of the cog-wheel of administrative machinery [which] worked so smoothly and surely'.[34] In fact photography was most extensively employed as 'a printing-press in the field', accurately and rapidly compiling and reproducing maps, plans and sketches of routes and reconnaissance. In this way, it was literally part of the engineering of the campaign.

The Royal Engineers also 'photographed all important and interesting scenes' on the route to Magdala.[35] They frequently presented a picture of a carefully engineered and well-ordered campaign, apparent in scenes such as General Napier and his commanding officers scanning maps (see illus. 27). Under the supreme command of General Sir Robert Napier of the Royal Engineers the campaign was certainly planned and executed with technical precision. Histories of the war read like accounts of a grand construction project, which in many ways it was. Engineering feats, accomplished with the help of Indian troop labour, included the building of piers at Annesley Bay and a

28 Royal Engineers, 'Camp at Zoola 1868'.

29 Royal Engineers, 'Balooch Regiment', 1868.

coastal base at Zoola, a railway line ten and a half miles inland involving eight bridges, as well as numerous roads, wells and pumps. The sheer scale of the Abyssinia Campaign is captured dramatically in photographic panoramas of the landing site and principal camps of the expedition. The panoramic view of the base camp at Zoola (illus. 28) shows the vast technical machinery, from the railway to the stacks of camp equipment, on which the expedition depended. Such panoramas provide grand, sweeping views of a landscape disciplined by British engineering and military might.

Panoramic views of military campaigns were hardly novelties and photography itself had been used in the making of panoramas since the 1840s. Roger Fenton made photographic panoramas of the plateau of Sebastopol and of Balaclava during the war in the Crimea in 1854–5. Specialized cameras, such as John Robert Johnson's Pantoscope (1862), had even been devised for the purpose. The Royal Engineers constructed their panoramas in the traditional way: through a series of individual, overlapping views.[36] Such panoramas were by no means easy to construct, demanding a series of even exposures from a level camera on a steady tripod, the camera being rotated a precise distance for each view. Because each segment required a different plate to be prepared, exposed and developed in turn, the resulting overall view was by no means as instantaneous as it appeared. This is demonstrated in illus. 28, where at least one person appears twice.

Such panoramas were designed to show the scope and order of the campaign to a home audience. In the field too, such technological sophistication had symbolic as well as practical uses. Sir Stafford Northcote, the Secretary of State for India, claimed that engineering technology, notably the equipment to convert sea water into drinking water, 'enabled us to impress upon some of the native chiefs and their representatives an idea of our skill and power'.[37] Indeed, the campaign was considered by many to have been largely an 'engineer's war', an instance of a larger process through which 'war [was] becoming more a question of science'.[38] The precision of the campaign is also registered in the sense of order which pervades photographs of imperial troops, such as 'Balooch Regiment' (illus. 29). Here the expeditionary force dominates a landscape otherwise devoid of human presence. The neat rows of white tents of the camp are complemented by the rows of soldiers of the 27th Bombay Native Infantry. These grids of men and tents, and the discipline and organizational strength they symbolize, impose a semblance of visual order on to a rugged Abyssinian landscape.

Unlike commercial photographers who accompanied earlier and subsequent campaigns, the photographers of the Royal Engineers were not treated as privileged artists. Nor were they individually acknowledged on their photographs. Their work was represented as a collective record rather than a series of subjective studies. Nevertheless, they did have something in common with professional photographers and painters in that they were still largely operating within the constraints of an aesthetic which represented war as a noble, disciplined pursuit.

The selection of appropriate subjects for photography was not simply a matter for the individual photographers, who were under orders from staff officers. Major Pritchard, who commanded the 10th Company, 'directed the photographers' and was apparently 'indefatigable in his endeavours to obtain interesting subjects for the camera'.[39] General Napier himself was said to have requested that a photograph be taken of the dead King Theodore. Owing to a delay this request was never carried out, to the disappointment of one commentator, who noted: 'How eagerly should we all have scanned the portrait here, in England, in endeavouring to read from the lines and markings of the brow the character of the determined warrior we had vanquished!'[40]

Military photographers were by no means uninterested in capturing picturesque subjects, as many of the Abyssinian photographs show. For a start, the training of Royal Engineers in photography was not concerned exclusively with technical expertise. As William Abney pointed out in his influential instruction book on photography, originally drawn up for use at the School of Military Education, 'to become a good photographer ... it is necessary to turn to it with an artistic and scientific mind'.[41] Hence Abney recommended that military photographers should, in addition to learning the technical aspects of photography, study guides such as *Pictorial Effect of Photography* (1869) by H. P. Robinson, with whom Abney later collaborated in a book on *The Art and Practice of Silver Printing* (1881). However, the aesthetic ambitions of commanding officers were not necessarily matched by an adequate understanding of photography's technical limitations. One commentator regretted that much useless work was done in Abyssinia because staff officers were ignorant of photography:

sometimes the mules had to be halted and the boxes unpacked during a long march in a drizzling rain in order that a picture might be attempted of some mountain or other, the top of which was enveloped in a dense fog, simply because a staff officer had expressed himself to the effect that the whole would make a grand picture.[42]

As well as having to deal with the aesthetic inclinations of officers, the Royal Engineer photographers had to contend with arduous environmental conditions, where high temperatures, frequent shortages of water and excessive dust made photography extremely difficult. Nevertheless, the surviving images are of a technical quality and aesthetic consistency comparable with the output of professional photographers. The photographs and the chief photographer, Sergeant Harrold, were duly acclaimed in Britain following the campaign. At a meeting of the London Photographic Society in December 1868, H. B. Pritchard praised the photographers for providing 'a clear conception of the nature of the country, for presenting to our view the various difficulties met with during the army's pilgrimage to release the prisoners, for giving us a true picture of King Theodore's stronghold'.[43]

Thus the application of photography to record scenes of the campaign was just as important as its function as a technique for reproducing maps and field sketches. As Pritchard's statement shows, the Abyssinian photographs were held to reveal the 'nature of the country' in full clarity. To be sure, the photographs represent more than simply impartial records or picturesque witnesses to the campaign; they should also be understood as part of a discourse of geographical science.

Science and Warfare

The Abyssinia Campaign was not only a military expedition, but was also an extensive scientific enterprise. As well as correspondents for major newspapers, the expedition was accompanied by official representatives from major scientific institutions, including the Geological Survey of India, the Zoological Society, the British Museum and the RGS. The expedition was particularly significant for the RGS whose then President, Sir Roderick Murchison, wielded his considerable influence to ensure that it was fully involved in the planning and completion of the expedition. For Murchison, the Abyssinia Campaign was a perfect means of advancing and proving the imperial utility of geographical science. Such views had a favourable reception at the RGS, an institution whose largest single group of Fellows, officers and council members throughout the Victorian period was drawn from the upper ranks of the army and navy.[44]

To many Victorian geographers and soldiers, war and geography were two sides of the same coin. Just as geographical knowledge had a major influence on military campaigns, war and its strategic planning made considerable contributions to geographical science. Military figures such as Major Charles W. Wilson, director of the Topographical Department of the War Office, therefore argued for 'the study of geography as a branch of military science' and for greater encouragement to be given to 'officers in our foreign possessions and colonies . . . to engage in geographical research'.[45]

Since 'knowledge is power' and almost any country in the world was a potential 'theatre of war', Wilson saw it as imperative that an effective imperial government should acquire as much geographical information about foreign territories as possible. This was more especially the case, he argued,'with regard to the little known districts, inhabited by uncivilized or but partially civilized races, that lie beyond the frontiers of many of our foreign possessions and colonies'.[46]

The RGS's library and map room was thus widely recognized as a highly important source of military intelligence.[47] As one commentator put it in 1874:

> the military and civil servants of Her Majesty well appreciate the value of the Society's map room. No sooner does a squabble occur – in Ashanti, Abyssinia or Atchin – than government departments make a rush to Savile Row and lay hands on all matter relating to that portion of the globe which happens to be interesting for the moment.[48]

The RGS had begun compiling and acquiring information on Abyssinia as soon as preparations for an expedition started, adding collections of books and views relating to Abyssinia to its map room and library collections, all of which were made available to War Office departments.[49] While the Army Medical Department compiled information regarding the climate and diseases of the region, Lieutenant-Colonel A. C. Cooke of the Topographical Department, under its director, Sir Henry James, amassed geographical information on the field of exploration. Cooke used the RGS's resources, including illustrated travel books, plates of natural history, atlases and maps in compiling his *Routes in Abyssinia* (1867), a report on the terrain and potential expeditionary routes.[50] In return, the Topographical Department supplied the RGS with copies of the textual information it compiled and views it reproduced by both lithography and photography.[51]

Naval and army officers had long produced 'views' made during overseas expeditions and campaigns, including the Niger Expedition of 1832–3,[52] and the Crimean War of 1853–6.[53] Such 'views' often combined maps and pictures in the representation of strategic places. Similar views of Abyssinia were used to provide a pictorial reconnaissance before the campaign. James Ferguson of the Topographical Department of the War Office produced a series of twelve numbered lithographs of 'views' which were published along with a general map of Abyssinia and accompanying descriptive information.[54] The views themselves were reproduced from earlier European explorations, including those undertaken by Henry Salt for Lord Valentia (1802–6), and those by the Scientific Commission of the French government under Lieutenant Lefebvre (1839–43). It was from the latter, for example, that 'View of the Valley of Aouadi and of the Desert of Adal at the Foot of the Mountains of Ifat' was reproduced (illus. 30). An accompanying text noted that it 'represents the view of the great plains of Africa, as seen from the Abyssinian Highlands' and added Lieutenant Lefebvre's own geographical notes and descriptions of the scene. Linked to a general map of Abyssinia and texts of explorers, these views, often picturesque in style, were used to provide information on the religious architecture, topography, vegetation, game and climate of Abyssinia.[55] As documents of strategic significance, the views also supplied knowledge of possible military routes and even the potential military opposition which might be found in Abyssinia. Circulated within military and geographical in-

30 Anon., 'View of the Valley of Aouadi and of the Desert of Adal at the Foot of the Mountains of Ifat', in J. Ferguson, *Views in Abyssinia* (1867).

stitutions, such views were considered, along with maps and travel narratives, an important source of strategic knowledge to aid the campaign in Abyssinia.

The uses of such views in providing strategic information for military campaigning also helped promote the military applications of photography. This was the case, for example, in use of photography at the Ordnance Survey promoted by Henry James and Roderick Murchison in 1858–9.[56] It was also important in stimulating Francis and Robert Galton to put forward their 1865 system of using photography to make portable, stereoscopic maps of mountain terrain for use by military commanders.[57] Thus when Major Charles W. Wilson set about improving the Topographical Department of the War Office (1869–77) and establishing the Intelligence Branch of the War Office (1871) of which he was Chief, he was well placed to promote the use of photography as a form of strategic knowledge. Indeed, in 1870, as secretary of a committee set up by the Secretary of State for War, Wilson successfully recommended that the Topographical Department make a collection of 'photographs of the colonies and foreign countries' in addition to its collections of maps.[58] This recommendation was undoubtedly related to the efforts of the Topographical Department in the production of the official albums from the Royal Engineers' photographs of the Abyssinian Campaign. By distributing photographic albums to influential government, military and scientific institutions the War Office was further acknowledging the close reciprocal relationship between the military and geographical science, as well as the significance of photography in the representation of the theatre of conflict.

The RGS was not the only learned society to be of use in the planning of the campaign. The collections of the Asiatic Society and Government Records in Bombay were also put at the disposal of military departments. Nevertheless, the Abyssinia Campaign considerably reinforced the official and unofficial links between the RGS and the War Office. As well as being closely involved in the planning of the campaign, scientific institutions were also involved in its execution. Sir Roderick Murchison, who had promoted the exploration of Abyssinia ever since his first presidency in 1844–5,[59] was quick to bring to 'the attention of Her Majesty's Government ... the desirability of sending out some men of science with the military forces about to proceed to Abyssinia'.[60]

Although a similar request at the time of the Crimean War had been overlooked, on this occasion government interest was forthcoming and Clements Markham, then Secretary of the RGS, was granted leave of absence from his duties at the India Office to take up the official post of 'Geographer of the Expedition', alongside other official scientists, including a zoologist, an archaeologist and an antiquarian. The work of these official scientists was tied closely to the progress of the military forces. In this capacity as official geographer, Markham compiled geographical reports on the climate, topography, botany and geology of Abyssinia, as well as producing an account of the expedition.[61]

Markham accompanied the storming party as it entered Magdala and was reputedly one of the first to find the dead body of Emperor Theodore.[62] His activities on the expedition would thus appear to represent simultaneously a search for scientific knowledge and an enthusiastic participation in military victory. Moreover, as part of a major military expedition, his geographical survey was itself a form of conquest, with scientific knowledge being the prize.

With its army of scientists, the Abyssinia Campaign continued a tradition within European imperial military activity initiated by the Napoleonic invasion of Egypt in 1798. Napoleon's celebrated expedition involved numerous scholars, chiefly from his Institut d'Égypte, whose practical skills and specialist knowledge were put to political use. The resulting massive *Description de l'Égypte*, published in twenty-three large volumes between 1809 and 1828, was nothing less than an encyclopedic textual appropriation of Egypt, setting the terms of the modern relations of dominance between Europe and the Near East within the discourse of Orientalism. The Abyssinia Campaign similarly relied upon and produced various forms of scientific texts, from maps to photographs, whose claim to capture the actual betrayed the 'textual attitude' adopted by Europeans in their relations with non-European places and people.[63]

Views of Abyssinia

In common with many imperial campaigns of the Victorian era, the Abyssinian expedition sought moral sanction by adopting the tone of a holy

31 Royal Engineers, 'Addigerat Church', 1868.

crusade, representing itself as a 'mission' or 'pilgrimage' to rescue European hostages (including a group of missionaries) from an apparently merciless African emperor.[64] Unusually, however, Britain's Abyssinian enemies were also Christian. Indeed Abyssinia had long appealed to the European imagination, especially though the myth of Prester John, as a lost, ancient Christian kingdom. Those participating in the representation of the expedition were no less interested in Abyssinia's Christian heritage, as evidenced by the number of photographs made of churches, such as 'Addigerat Church' (illus. 31). Photographs were also made of Theodore's Christian artefacts, including Bibles and a cross.

Nevertheless, the presence of such signs of Christian culture in Abyssinia did not sit comfortably with the predominant contemporary image of Africa and Africans as lacking both history and civilization. For many European scholars the paradox was resolved by identifying such artefacts as relics of a long-lost, foreign civilization. Markham himself portrayed what he called the 'present barbarism' of Abyssinia as having resulted from its geographical isolation from the rest of Christendom and external 'civilizing influences'.[65] Thus he claimed that the masonry of a ruined church near Agila, of which he made a ground plan, 'is very far superior to anything the Abyssinians could build now, and must have been the work of foreign artists'.[66] Abyssinia, like other earlier destinations of European imperial expeditions and scholarship such as Egypt, was valued for its supposed links to a distant European past.[67] Similarly, Dr Beke, a well-known explorer of Abyssinia, emphasized the ancient and foreign origins of much Abyssinian architecture.[68] The churches and biblical manuscripts photographed by the Royal Engineers were consequently represented as relics of a decayed Christian civilization. Although the expeditionary scholars held out the possibility that this civilization was renewable, Christian artefacts were frequently represented as out of place in a barbarous land. The British therefore had little hesitation in appropriating Theodore's religious collections and royal possessions upon the capture of Magdala, when the plunder from the city was auctioned to raise money for the troops. Theodore's collection of religious books and manuscripts were acquired by Dr Richard Holmes for the British Museum, while his royal seal and personal locket, 'found' by Markham in Theodore's house, eventually found their way to the RGS, where they remain to this day.[69]

The projection of Abyssinia as merely an exotic backdrop for a British narrative of national and imperial prowess, Christian duty and civilization is also apparent in other photographs. The panorama of the Abyssinian expedition camp at Senafé (illus. 32) shows clusters of white tents and rows of equipment set out on an undulating stony plain. As in the other panoramas and many of the views, the photographer has chosen a high vantage point from which to survey the surrounding scene. Despite the lofty position of the observer the landscape ends abruptly against a range of large rocks. With its wide scope and abrupt horizon the view has much of the artificiality of a theatre scene or painted backdrop. Echoes of the technologies of visual entertainment are less surprising if we consider the popular appeal of both exotic scenery and military action in spectacular show panoramas which were beginning to be revived in European capitals.[70] The Royal Engineers, in making such panoramas, and the Topographical Department of the War Office, in distributing and displaying them, were thus following a long tradition of spectacular view-making. Indeed photographic panoramas of Abyssinia, exhibited at the Topographical Department or the RGS, might be thought of as the scientific equivalents of the large dioramas of 'The Death of Theodore' which were exhibited in 1868 at the Royal Polytechnic twice a day to enthusiastic audiences.[71]

While their raised perspective and all-embracing reach suggest an association between these photographs and other forms of panoramic spectacle, they also represent a form of cartographic surveying. Since the mid-1850s researches into the use of photography as an instrument for topographic surveying and map-making had been undertaken in Europe, particularly through the experiments of Colonel Aime Laussedat, the French Battalion Chief of Engineers. However, until the 1870s the expense and bulk of the apparatus proved largely prohibitive to its widespread use, particularly on

32 Royal Engineers, 'Camp at Senafé 1868'.

military campaigns. Nevertheless, the Royal Engineers had for some time advocated photography as a scientific means of surveying landscape. In 1860 Captain Henry Schaw, head instructor at the School of Photography and Chemistry at Chatham, noted:

> in surveying boundaries of different countries, photographs of remarkable natural features of the country, which may either occur in the boundary line or be visible from certain points in it, will tend to fix the positions of the line with great certainty.[72]

In response to such calls, photography had already been employed by the Royal Engineers in surveying activities in various parts of the world, most notably during the North American Boundary Commission Survey (1858–62) which delimited the boundary along the 49th parallel between the United States and the new British colony of British Columbia. Following the desertion of the first sapper photographer in 1859, two more were sent from England with Lieutenant Samuel Anderson – who, as Assistant Instructor in Photography at Chatham in 1867, played a key role in organizing the photographic equipment for the 10th Company Royal Engineers in Abyssinia.

The commission included photographs as part of its report to the Foreign Office, a set of eighty-one official photographs being sent to the Secretary of State for Foreign Affairs in early 1863.[73] Photographs of settlements, survey camps, boundary markers and cuttings through forests were thus used partly to represent the obstacles faced by the party and to witness their progress. However, the photographers worked in difficult conditions and their bulky apparatus was not well suited to operations in the field. That they persisted, despite the difficulties, shows that photography was employed not merely as a convenient means of commemorating the survey. Indeed, photography was itself a technique of visual survey. In the attempt to provide a map of 'views' of this colonial territory the Royal Engineer photographers worked alongside a team of observers which included astronomers, a naturalist, a botanist and a geologist. Like the resulting scientific texts and survey maps, their

33 R. Baigre, 'Encampment at "Wah" or "Weah", Officers Surveying the Terrain', 1867, photograph of watercolour sketch.

photographs were intended to develop scientific knowledge of the landscape and 'Indian tribes' surveyed.[74] By the time of the Abyssinia Campaign, therefore, the Royal Engineers were well experienced in using photography in the course of their survey work.

The use of photography to view the landscape was also closely linked to an older tradition of pictorial 'views' and cartography. 'Landscape views', particularly as delineated in pencil sketches and watercolours, had long been used in military and exploratory practice as forms of topographical and reconnaissance surveys.[75] Even before the advent of photography pictorial 'views' had been used, in association with maps, as a form of survey. W. Allen's *Picturesque Views on the River Niger* (1840), for example, included a three-part panoramic view 'taken from the summit of a little hill' which mapped out the territory below.[76] Similarly, in Abyssinia sketches and watercolours were used to provide convenient views of the topography. 'Encampment at "Wah" or "Weah", Officers Surveying the Terrain' (illus. 33) is one of a number of watercolour sketches made by Major R. Baigre of the Bombay Army, whilst engaged with the advance reconnaissance party in November 1867. As well as showing officers surveying the terrain, the picture is annotated with observations concerning routes and resources, such as sites of 'good water'. The

success of the Royal Engineers at football and cricket disproved such theories.[92] Sir Robert Napier's organization of altogether more extensive team sports did much to heighten the status of the 'scientific soldiers'. It certainly made a public hero of Napier, who earned a peerage and immortalization in a bronze statue with a commanding view of Hyde Park in London.

General Napier was also made an Honorary Fellow of the RGS, and Sir Roderick Murchison was delighted to be able to claim in 1868 that 'the distinguished General who has accomplished these glorious results is a man of science, and is particularly well versed in Geography'.[93] Murchison's delight was widely shared in scientific circles, since military victory in Abyssinia was widely recognized as simultaneously a triumph for British geography. Hearing the news of the fall of Magdala, the president and council of the Berlin Geographical Society sent a telegram congratulating the RGS on 'this new success of British valour, benefiting geographical science'.[94] Scientific institutions had much to gain from such exercises, for just as warfare became more 'scientific', science absorbed the ethos of military campaigning.

For scientific institutions such as the RGS, the Abyssinia Campaign represented a grand opportunity to display both the practical usefulness of science in the exercise of imperial military authority and the importance of military expeditions in the production of geographical knowledge. In May 1868, approximately six weeks after the storming of Magdala, Murchison proudly asked the assembled RGS Fellows:

> When has Europe marched a scientifically organized army into an unknown intertropical region, and urged it forward as we have done, for hundreds of miles over chain after chain of Alps amid the grandest scenery? And all to punish a dark king, of whom we only know that he was an able but unscrupulous tyrant who insulted us by unjustly imprisoning our countrymen. This truly is a fine moral lesson which we have read to the world; and as, in addition, we reap good scientific data, the Abyssinian Expedition will be chronicled in the pages of history as more worthy of an admiring posterity than many a campaign in which greater political results have been obtained, after much bloodshed, but without the smallest addition to human knowledge.[95]

Justification on moral and scientific grounds was undoubtedly made easier because this was a campaign set in Africa and posed as much against the natural world as against human enemies. For this reason Murchison, like John Ruskin, found 'small wars' far easier to sanction than European ones.[96]

As one of many 'small wars' fought in Africa throughout the nineteenth century, the Abyssinia Campaign anticipated Britain's subsequent major role in African colonization. Indeed, more permanent forms of colonial influence were conceived and debated at institutions such as the RGS.[97] The explorer Samuel Baker advocated annexation, arguing that northern and western Abyssinia 'might be made one of the finest cotton producing countries in the world'. John Crawfurd, on the other hand, rejected Baker's suggestion, since

36 W. E. Fry, 'Electric Light at Macloutsie Camp', *Occupation of Mashonaland* (1890).

'the inhabitants of Abyssinia were barbarians, and no barbarians ever did produce cotton'. Such blatant political theorizing was usually dampened by the RGS's leadership, and on this occasion Stafford Northcote, the Secretary of State for India, felt it necessary to repeat that the official aim of the expedition was simply to rescue Theodore's captives and that 'no other consequences would follow'.[98] Indeed, no direct British territorial gain resulted from the Abyssinia Campaign and it preceded by several years Europe's well-known 'scramble' for African territory (1876–1912).[99] Yet contemporary geographical discourse shows how British imperial influence in Africa in the late 1860s was promoted no less powerfully for being pursued through the teaching of 'a fine moral lesson' and the reaping of 'good scientific data' than through the acquisition of territory by itself.

To the British public and to military and scientific circles, then, the Abyssinia Campaign represented an adventurous exploration into the unknown. Its significance in the present context lies in what it reveals of the place of photography and science within the art of campaigning as it was conceived in the second half of the nineteenth century.

In their use of photography as a form of regulated technology, the Royal Engineers were also previewing the increasing association of photography with forms of technological warfare. This is shown, for example, in William Ellerton Fry's photographs of the occupation of Mashonaland in 1890. As well as being intelligence officer, meteorologist and assistant to Frederick Selous on the Mashonaland Pioneer Column in 1890, Fry was official photographer, producing an album representing this imperial expedition.[100] The term 'occupation' is apt in describing what was a quasi-military operation. Some 200 'pioneers', attracted by rewards of large estates and free gold claims, accompanied by 500 police from the newly formed British South Africa Company and many hundreds of Africans, embarked northwards to break a pathway into Mashonaland. Fry's photographs show the leaders and officers of the Pioneer Column and portray the strength of their expedition.[101] Scenes of camps are reminiscent of the Abyssinia photographs in their picturing of an orderly force in an unknown landscape.[102] The massive African labour force which was used to forge the route was, by contrast, given almost no attention. Among Fry's photographs are views of the vast, steam-powered electric light which the force used to light up the country at night (illus. 36). Light was thus used as a weapon to expose the landscape and potential enemies to the vision of the colonial forces. As I go on to show in the following chapter, the association between the camera and weapons of war found particular expression in the practices of hunting on the colonial frontier.

As a 'scientific' enterprise, the Abyssinia Campaign also shows how photography fitted into an established tradition of European imperial military representation in which the texts of science were significant forms of cultural plunder. For, as the geographer Markham noted, 'The men of science who accompanied the expedition have not returned empty-handed.'[103] To be sure, artillery was a more effective means of domination in the field than any camera. Yet, as a supposedly immediate and accurate visual record, photography also conquered the geography of Abyssinia, bringing visual residues of its landscapes back home as trophies.

As I have noted, the power of photography as a witness of war had been recognized since before the Abyssinia Campaign. Just as with Roger Fenton's photographs of the Crimea fifteen years earlier, the selective and conventional imagery of military might presented by the Royal Engineers' photographs of the Abyssinia Campaign resulted in part from the consciously controlled nature of photographic activity. For a start, no photographers besides the Royal Engineers accompanied the expedition. The photographic representation of the campaign from this single point of view could then be further controlled, both in the construction of individual photographs and in their incorporation into an official record by the War Office. As cameras became both more portable and more accessible to ordinary soldiers, official control of the making and use of photographs of war became an increasingly complex operation. By the time of the South African War of 1899–1902 a range of photographic images, notably those made by ordinary soldiers, was revealing very different views of the 'realities' of colonial warfare.[104]

4 Hunting with the Camera

In twentieth-century Western society, where taking photographs consists of 'loading', 'aiming' and 'shooting', the camera has become what Susan Sontag has described as a 'sublimation of the gun'.[1] Yet this process of sublimation was well under way in the second half of the nineteenth century, most especially within the language and practices of Victorian hunting. From the late 1850s explorers, soldiers, administrators and professional hunters began to employ the camera to record images of dead animals for purposes of scientific documentation and as evidence of their hunting achievements. Yet despite the wealth of this photographic record, in published accounts, official records and private albums, the place of photography within the ritual of colonial hunting has barely been considered. In exploring the aetiology of the camera/gun analogy in more detail within Victorian discourses of hunting I will be forced to touch on a range of broader practices than an initial definition of hunting might at first suggest, including natural history, exploration, mountaineering and conservation.

The colonial hunter was one of the most striking figures of the Victorian and Edwardian imperial landscape. Frequently pictured posed with a gun beside his recently killed prey, or surrounded by the skins, tusks and other trophies of an expedition, the hunter is, to present-day eyes, the archetypal colonial figure. The most famous big-game hunters are still remembered, variously, as heroic adventurers, intrepid explorers, accomplished naturalists and, perhaps ironically, pioneers of conservation.[2] Yet hunting was not limited to those who made it a full-time career. Many British colonial administrators, soldiers, settlers and travellers participated in the chase and killing of wild animals, both as a form of sport and as a scientific pursuit. Interest in pursuing zoological 'specimens' for private and national collections was fostered in part by the dramatic upsurge in the popularity of natural history in the first half of Queen Victoria's reign. Encouraged by the commonly held faith in natural theology, the paraphernalia of rational amusement and scientific instruction – from microscopes to natural history books – abounded in what has been described as the 'heyday of natural history'.[3] Hunting men were also inspired by the proliferation of popular literature and images of hunting which frequently pictured the hunter as a manly adventurer and hero of Empire.

The varied interests and activities of Victorian and Edwardian hunters certainly need to be set within the wider context of European imperialism and the relations between human groups, animals and the environment.[4] As Harriet Ritvo has argued, the hunting, collection and display of wild animals were intimately associated with the ideology of Empire.[5] In a major study of hunting, conservation and British imperialism, John MacKenzie has also shown how the techniques and ethos of European hunting were embedded in the imperial enterprise.[6] Other historians have echoed these approaches, arguing that hunting played a highly significant role within colonial expatriate cultures.[7]

Even before the development of photography, hunters relied on visual techniques to record their experiences. Captain William Cornwallis Harris, for example, one of the earliest and most famous European hunters in Southern Africa (in 1835–7), used his detailed drawings of hunting scenes, indigenous peoples and animals to illustrate his highly popular account *The Wild Sports of Southern Africa* (1838).[8] Harris's drawings of animals, set within naturalistic landscapes devoid of human presence, also received specialist attention in his *Portraits of the Game and Wild Animals of Southern Africa* (1840).[9] While some commentators predicted that the unreliable and occasionally caricatured creations of the artist would be replaced by truthful delineation of nature in photography, it was soon realized that the sketchpad had many advantages, not least versatility and practicality, which would ensure its lasting role in the depiction of flora and fauna.

Nevertheless, as soon as hunting became a suitable activity to commemorate in photographs, hunts began to be undertaken primarily for purposes of photography. In the early 1870s, for example, the amateur photographers W. W. Hooper and V. S. G. Western made a series of twelve photographs which were offered for sale under the general title *Tiger Shooting*. The carefully staged photographs recreated for the viewer the principal events of a tiger hunt, from the colonial hunters' first camp to the tracking and shooting of the tiger and its skinning.[10] 'Bagged' (illus. 37) reconstructs the moment at which the colonial hunter, atop a large rock, has finally claimed his victim. Such photographic imagery, as I go on to show, was as much about capturing moments in time as capturing nature. Indeed, during his varied military career in India from the late 1850s to the late 1880s, Willoughby Wallace Hooper (1837–1912) took his passion for precision-timing photography to extreme lengths.[11] For example, in the 1880s he was nearly court-martialled for his attempts to secure photographs of the British firing squad execution of dacoits in Burma by timing the release of his camera shutter to the exact moment when the order to fire was given. Hooper's arrangements caused several delays to the proceedings and, in the minds of some, unnecessary torment to the convicted

37 W. W. Hooper and V. S. G. Western, 'Bagged', in *Tiger Shooting* (*c.* 1870).

men. Hooper's case is undoubtedly extreme. However it indicates both the close association of shooting with guns and cameras and the fascination early photography held for its seeming ability to fix moments in time ordinarily invisible to the naked eye. In particular, it demonstrates the operation of cameras as what Roland Barthes called 'clocks for seeing' and photography's intimate relation to death in its certification of a presence 'that has been'.[12]

Hooper and Western's series of photographs of *Tiger Shooting* was also informed by the widespread fashion for sporting painting in this period. While images of hunting and of animals were a well-established tradition in European art, the school of English sporting painting reached the height of its popularity in the Victorian era, as demonstrated by the work Sir Edwin Landseer (1802–73). Through royal patronage and mass-produced engravings of his work Landseer became one of the most popular artists of the time.[13] His hunting scenes, such as *Royal Sports on Hill and Loch* (1850), which portrays the Royal Family's sporting recreation in the Scottish High-

38 'Their Excellencies Just After Shooting', Lala Din Diyal & Sons, *Souvenir of the Visit of H. E. Lord Curzon of Kedleston, Viceroy of India to H. H. the Nizam's Dominions, April 1902.*

lands, evoke strongly the Victorian faith in the authority of human groups over a tamed and improved natural world and more generally reflect the significance of this 'sport of kings' within Victorian culture.[14]

The same iconography of dignitaries dominating the natural world is apparent in photographs commissioned to record official sporting tours. 'Their Excellencies Just After Shooting' (illus. 38), for example, is one of a series of photographs of a tiger hunt included in an official souvenir album by the commercial photographers Din Diyal & Sons commemorating the visit of Lord Curzon, Viceroy of India (1899–1905), to the Nizam of Hyderabad in 1902. With the spread of British rule and the extension of the railway, the large, organized *shikar* (tiger hunt) had become an important way for Indian rajas to entertain the Queen's representative.[15] Many years later Curzon reflected, 'The sporting tours of the modern Viceroy, with their wonderful adventures and often prolific results, are an experience which no one who has enjoyed them would willingly have missed, or can ever forget.'[16]

By the turn of the century, photography had become an indispensable part of the ritual of these sporting tours in India. By 1902 Lala Din Diyal (1844–1910) was an accomplished commercial photographer with established studios in Secunderabad and Bombay. Well known as a state photographer, working for earlier Viceroys such as Lord Northbrook (in 1872–76) and Lord Dufferin (in 1884–88), as well as the official photographer to the Nizam of Hyderabad from 1884, Diyal was well practised in the art of commemorating the authority of his patrons.

For Lord Curzon such photographs were treasured personal mementoes of his hunting achievements. Indeed, in his album this photograph has been further annotated with the note 'Tiger shot dead through back of head at 70 yards'. In addition, the evidential authority of photography could be exploited within a conventional iconography – derived largely from painting – of human dominance over the natural world. Standing nearest the photographer at the head of the slumped tiger, clutching his gun, Curzon adopts the conventional stance of the victorious huntsman and landowner. His confident pose symbolizes British authority over India at the moment when Britain's Empire was at its zenith. Landseer himself would have been impressed by such a striking photographic representation of the assertion of human and imperial power ritually enacted through the colonial hunt. As Landseer's work shows, the iconography of humans, particularly monarchs, laying claim to landscape and animals as property was well established. A similar social hierarchy is here re-enacted as Curzon claims his tiger and the beaters and servants recede into the dark undergrowth in the background. Yet as official photographer to the Nizam of Hyderabad – shown prominently in this image on Curzon's left – Diyal's commemorative album of Curzon's official visit was also representing the Nizam's own local ruling authority. Many photographs in the *Souvenir* show more clearly the vast resources put at the disposal of the Viceroy and Vicereine on their official *shikar*.[17] Indeed, it was from Indian rulers that the British had appropriated the symbolic associations of the *shikar*. Moreover, it was only a partial appropriation as the symbolism was jointly shared by English and Indian rulers. Indeed, on Lord Curzon's first Viceregal tour in India in 1899 Diyal had photographed the first tiger shot by Curzon in India with the Viceroy and the Maharaja Sindhia of Gwalior each resting one foot on the dead animal.[18]

Tigers had long been treated as symbols of power by Indian as well as British rulers. They evoked particular importance for Tipu Sultan of Mysore, who had their form used liberally as emblems on furniture, costumes and weapons. After Tipu was defeated and killed by the British in 1799, his mechanical model of a tiger killing an Englishman became one of the most famous London exhibits of the nineteenth century. Originally housed at the East India Company's Oriental Repository at Leadenhall Street, in 1880 the model was moved to the South Kensington Museum, where it remains.[19] The model had arguably been designed to celebrate an actual incident in 1792 when the son of Sir Hector Munro, British army general and arch-enemy of Tipu, had been killed by a tiger. Its capture by the British and subsequent display in London commemorated the British imperial victory over Tipu and the taking of Seringapatam, Mysore's capital, in 1799. Following the Mysore Wars the tiger became equated in the British imperial imagination with a specifically 'Oriental' ferocity and motiveless violence.[20] The association between controlling or killing tigers and the conquest of India and Indians remained a powerful one for the British throughout the nineteenth century.[21] Curzon was not unaware of such associations, taking much interest in the trophies of the Mysore Campaign, particularly Tipu's throne and gold tiger-head footstool, parts of which decorated Government House.[22]

Empire Building and Natural History

Animal symbolism was just one element in the attraction of hunting for Victorians. For men like Curzon, hunting embodied a particular mode of understanding the natural world and was part of his broader interests in both the natural sciences and the extension of Empire. In this he was by no means unique. Early commissioners in Africa, for example, such as Frederick Jackson, Harry Johnston and Robert Coryndon, were commonly as interested in natural history as in colonial government. Similarly, Sir Henry Bartle Edward Frere (1815–84), who was Chief Commissioner of Sind (1850–59) and Governor of Bombay (1862–6) before becoming Governor of Cape Colony in 1878, was an enthusiastic geographer, becoming, as Curzon also later would, President of the RGS.[23] The frontispiece of Sam Alexander's 1880 album *Photographic Scenery of South Africa* shows a vignette of Sir Bartle Frere – to whom the album was dedicated – surrounded by the natural 'scenery' of South Africa (illus. 39).[24]

Frere was himself interested in photography, acting as patron of the Bengal Photographic Society from 1859–62, while a member of the Viceroy's Council and Lord Canning's confidential adviser. His acquisition of collections of photographs during his imperial career in both India and Africa was part of his wider interest in the geography of the natural world.[25] With the photograph of Frere neatly framed by the flora and fauna of South Africa, including a noble Zulu warrior and principal colonial products represented by elephant tusks and the numbered bale of wool, this image is yet another variation on a theme of Britannia, with the fruits of the natural world being displayed for Britain's use (see illus. 5). Indeed, as a colonial administrator of the natural world it was Sir Bartle Frere, with the colonial power he represented, who was at the centre of the display. Alexander's volume of 100 photographic 'views' and introductory letterpress attempted to show 'in its proper light' the country where Sir Bartle Frere had 'done so much good

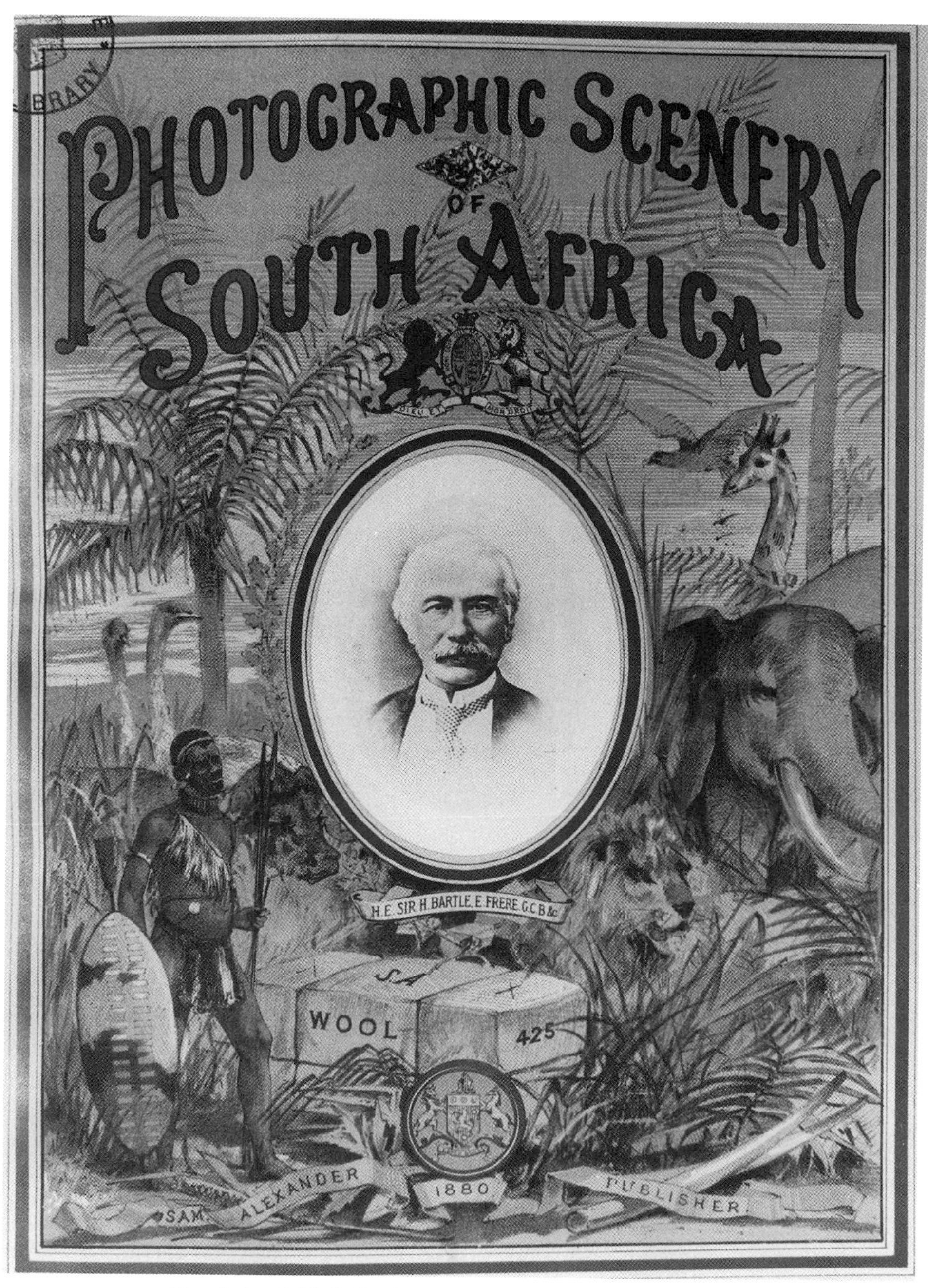

39 Frontispiece (with Sir Bartle Frere), Sam Alexander, *Photographic Scenery of South Africa* (1880).

work'.[26] Photographs of a natural world beginning to be reordered by European settlement, farming and industry are juxtaposed with those of a yet untamed nature where indigenous people are entwined within dense vegetation.[27] In short, the album told a triumphant story of British colonization and the gradual improvement of colonial administration and opportunity through the work of 'public institutions' and 'illustrious men' such as Sir Bartle Frere.

Many major figures of Victorian science were initiated into adulthood and the scientific community through hunting. Francis Galton, for example, after gaining his inheritance on the death of his father in 1844, devoted himself to sport in Britain for several years, an experience which prompted him to undertake a more extensive expedition in Africa in 1850.[28] Here Galton put his hunting to use not only to provide food but also in the symbolic rituals of the colonial encounter, as when he confronted a chief of the 'Namaquas' dressed in his fox-hunting kit and riding an ox.[29] Galton's companion, Charles Andersson, was an even more enthusiastic hunter and collector and his 1856 account of this expedition and his subsequent travels to Lake Ngami contained dramatic lithographs of hunting and wildlife scenes by the German artist Joseph Wolf.[30]

Galton considered hunting to be an essential art for travellers, soldiers and explorers; it occupies – in both practical and sporting dimensions – a central place in his famous *Art of Travel*.[31] First published in 1855, *The Art of Travel* remained popular with travellers and hunters for many years to follow. As late as 1893, the big-game hunter H. A. Bryden claimed it was invaluable for anyone wishing to 'exploit south Africa'.[32]

Explorers such as David Livingstone certainly had to learn the arts of hunting, both to feed expeditionary parties and for self-protection. While sport was not Livingstone's main concern, he travelled extensively with full-time hunters, such as William Cotton Oswell who hunted in southern Africa between 1844 and 1851. While such hunters associated themselves with explorers like Livingstone to enhance their own reputations, they were often praised by figures like Galton as contributors to scientific knowledge.[33] Accounts of hunting, such as Captain James Forsyth's *The Highlands of Central India* (1871), were often simultaneously presented as records of geography, natural history and anthropology.[34] Conversely, men whose photographs earned them reputations as anthropologists, such as H. W. Seton-Karr and M. W. Hilton Simpson, were also keen sportsmen.[35] Thus many of these scientific hunting men were tracking down knowledge as well as big game. Whether pursuing the source of a river or a rare specimen of game, such knowledge was frequently linked to the language and politics of imperial expansion.

Of all the big-game hunters and adventurers whose exploits were explicitly linked to the extension of Empire in Africa in the 1880s and 1890s, Frederick Courteney Selous was probably the best known. Selous spent some two decades (1872–92) hunting and Empire-building in Africa; activities which he publicized through his popularly acclaimed books and lectures.[36] In his *Travel and Adventure in South-East Africa* (1893), for instance, he showed how his six years' hunting for natural history specimens proved invaluable in his securing the route of the Mashonaland Pioneer Column in 1890. Indeed, it was the experience gained in years of travel and hunting which, Selous claimed in 1893, 'enabled me to play my part in the actual occupation of Mashunaland'.[37]

Like other natural history hunters, such as Chauncey Hugh Stigand and Denis Lyell, Selous claimed that excursions into the natural history of game elevated him above mere shooters.[38] Indeed, such hunters frequently justified their killing by claiming to be scholarly shooters collecting for scientific specimens. Selous collected extensively for natural history museums and, as his bronze statue in the Natural History Museum in South Kensington shows, he gained widespread recognition for his scientific services. In addition to popular adulation, therefore, Selous achieved praise at scientific institutions such as the RGS, where the results of his expeditions were often recounted and where he was commended for his observations on natural history, zoology and topography. While Selous used the RGS as a source of scientific respectability, the Society bestowed upon him grants and medals in the 1880s and 1890s, including its prestigious Founder's Medal for his exploration of Mashonaland and Matabeleland[39] in the 1880s and his work in leading the British South Africa Company's pioneer expedition to Mashonaland in 1890.

For men like Selous hunting not only provided useful knowledge; it symbolized crucial national characteristics. Thus he claimed, for example, that 'the expedition to and occupation of Mashunaland cannot but foster the love of adventure and enterprise, and tend to keep our national spirit young and vigorous'.[40] Such sentiments tapped widespread currents within Victorian society, where the white hunter and adventurer represented the type of energetic, pioneering white man upon whom Empire depended.[41] Like his onetime sponsor Cecil Rhodes, Selous used the supposed superiority of the white Anglo-Saxon race as justification for colonization in Africa. He drew on the same ideas of innate Anglo-Saxon racial characteristics to justify the South Africa Company's war against the Matabele (Ndebele) led by Lobengula, in which he took an active part, in 1893.[42]

While Selous was not himself an avid photographer, his illustrated publications and popular lantern-slide lectures show that he was clearly aware of the power of photographs as visual witnesses to imperial progress. For instance, it was at the instigation of Selous that William Ellerton Fry was employed as official photographer – as well as intelligence officer, meteorologist and assistant to Selous – on the Mashonaland Pioneer Column in 1890. The role of Selous in determining the kinds of photographs made for *Occupation of Mashonaland* (1890) is uncertain.[43] Yet Fry certainly included photographs of shot animals,[44] further showing how hunting was represented as an integral part of pioneering.

For many hunters photography was a convenient means of displaying their hunting and exploring achievements in lectures, books and exhibitions. Thus in February 1893 an exhibition at the RGS to coincide with a lecture by Selous displayed photographs, including a number by William Ellerton Fry, together with 'zoological specimens'.[45] Photographs were also part of the private collections of hunters. Among the personal belongings of Major Chauncey Hugh Stigand, another famous big-game hunter and natural historian,[46] for example, is to be found a small tobacco tin containing, along with lions' claws and bullets, a photograph showing an unidentified African servant holding a pair of tusks (illus. 40). By the time of his death in 1919 in action against the Dinka in southern Sudan, Major Stigand, then Governor of the Upper Nile Province, had built up a considerable reputation as a soldier, hunter, colonial administrator, and geographer, particularly in central and eastern Africa. His skills with a rifle and his knowledge of languages and natural history made him representative in the eyes of his contemporaries of an ideal frontiersman in Britain's tropical Empire. One obituary described him as 'one of a numerous band of self-reliant Englishmen for whom a distant and uncivilized frontier has a fatal attraction, and who do their best work where civilization and settlement are unknown'.[47]

Stigand's combined skills in hunting, natural history and soldiering gave him particular advantages when working in what he termed 'savage countries'. He expressed this best in his *Scouting and Reconnaissance in Savage Countries* (1907), one of a series of books he wrote on these subjects. Designed as a portable manual for the 'soldier scout', the book provided instruction on the 'strategy' and 'tactics' of campaigning, involving the 'tracking', 'stalking' and 'marking down' of both animals and human 'savages'.[48] The continuities between campaigning and hunting are continued in his *Hunting the Elephant in Africa* (1913), in which a chapter on 'Stalking the African' included his recollections of warfare against various human enemies.[49] Campaigning against Africans, like big-game hunting, was also measured as a form of entertainment. Stigand noted how in 'usual African warfare ... the show is hopelessly dull for the ninety-nine times and rather too exciting for the hundredth'.[50] Moreover, he considered both campaigns against 'savages' and big-game hunting to be useful in preparing British troops for 'real' warfare against 'civilized' enemies.[51]

In Stigand's work, sport was given a scientific as well as a military justification. His hunting manual, jointly authored with Denis Lyell, *Central African Game and Its Spoor* (1906), was designed as a guide to observing, tracking and shooting game. As Stigand put it, 'sport does not lie in the mere slaying, but in the patience and skill necessary to pick out and bring to bag a good specimen of

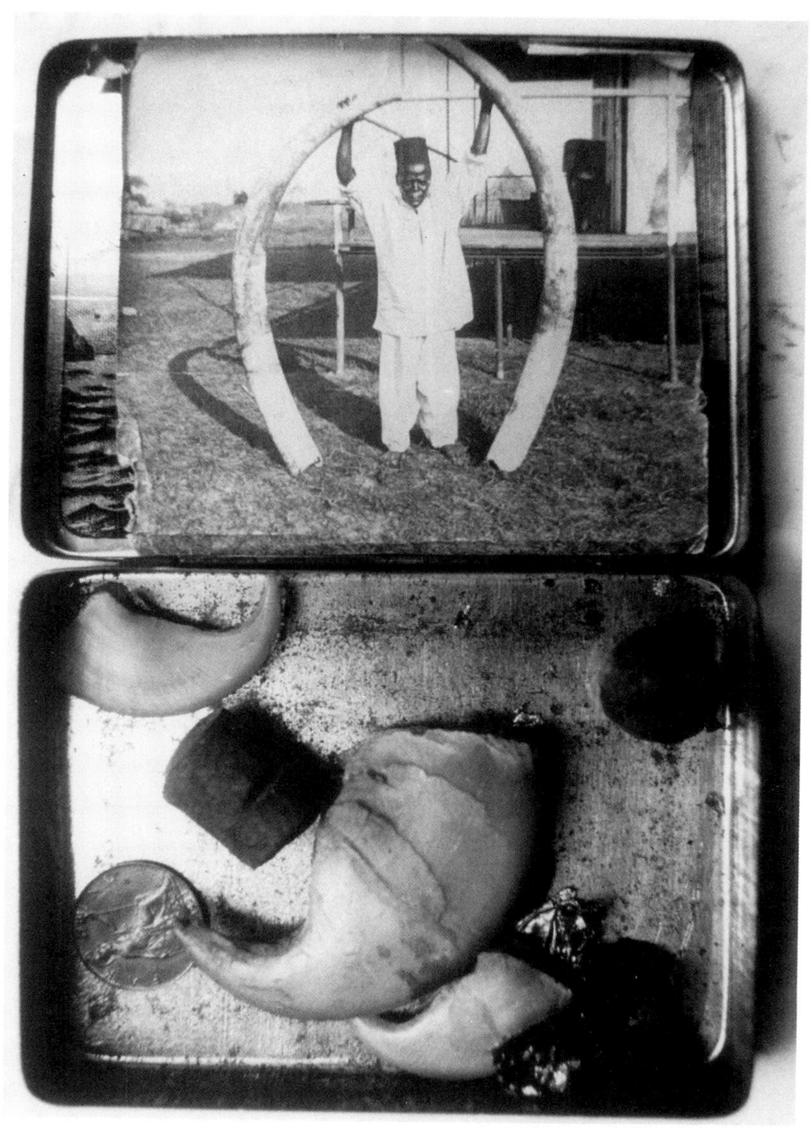

40 Photographs as personal hunting mementoes, C. H. Stigand Collection, Royal Geographical Society Archives.

its kind'.[52] Such rhetoric was certainly designed to enhance the status of the hunter. In his introduction to Stigand's *Hunting the Elephant*, Theodore Roosevelt commended big-game hunters 'who understand wild natives and animals'; those who, like Stigand, were not only adventurous hunters, but also 'excellent field naturalist(s)'.[53]

Hunting trophies, whether claws, bullets or photographs, also served as badges of manly identity, for hunting also embodied a romantic ideal of Victorian masculinity: independent, courageous, physically robust, honest and white. These assumptions are made clear in Stigand's unpublished book, *An African Hunter's Romance*, written while working in the Sudan. The book, which claimed to be a fictional plot interweaving real events and settings, is structured around the daring exploits of a big white hunter, notably in protecting womanly virtue or avenging its mistreatment.[54] This romance was more than mere fancy, as hunting was frequently characterized as an activity in which women were not fit to participate. From the late eighteenth century, women had been increasingly excluded from fox-hunting in Britain. This major rural pastime increasingly became a form of social bonding exclusively for men, its encouragement of manly virtues of courage and chivalry seen as a corrective to effeminacy.[55] The rise of big-game hunting on the outskirts of Empire, and its presentation as a manly occupation, might thus be partly seen as one response to the increasingly assertive visibility of particular groups of women in late Victorian society, notably the suffragettes.[56]

Many hunters celebrated the 'outdoor life of absolute freedom' as a refuge not only from modern, industrial Britain but also from women, or, as one African hunter put it, 'from the remembrance of "her" at home'.[57] A number of well-known hunters were strongly opposed to women's attempts to break from their conventional confinement within the domestic sphere. Even while absorbed in the loneliness of the African wild, Captain E. Lardner found time to voice his opposition to suffragettes.[58] The hunter, soldier and natural scientist Richard Meinertzhagen was also vocal in his dislike of suffragettes and resigned from the RGS over the hotly debated issue of women Fellows.[59]

Nevertheless, as the 'professional female globetrotters' who so appalled Lord Curzon were just then demonstrating, some women were exercising their ability to travel and to produce authentic knowledge.[60] Women travellers were certainly undertaking expeditions and recounting their experiences in ways that both accommodated and challenged the predominant male and imperial travel-writing ethos.[61] Mary Kingsley, for instance, adopted a distinctly unheroic stance in *Travels in West Africa* (1897) when she declared, 'I have seen at close quarters specimens of the most important big game of Central Africa, and, with the exception of snakes, I have run away from all of them.'[62]

Her almost proud admission of cowardice in the face of big game may be read as a sign of her generally ambivalent stance towards a masculine tradition of exploration or even a parody of male heroics.[63] For although she climbed mountains, fished and photographed, she had little interest in big-game hunting. However, a number of other British women were involved in hunting expeditions. Lady Margaret Loder accompanied her husband Reginald B. Loder on safaris to British East Africa in 1910–11 and 1912–13, shooting and photographing big game alongside him.[64] The Loders' safaris were organized by the tourist company Newland, Tarlton & Co. and were undertaken in some luxury, under the guidance of professional white hunters.

A more striking example is the hunting expedition privately undertaken and organized by cousins Agnes and Cecily Herbert in Somaliland in 1908.[65] In her account of the hunting trip, Agnes Herbert stressed that two women could undertake such a venture as well as any man. Indeed, the women competed with a rival trip organized by British men, whose attention and condescension they consistently rebuffed.[66] Agnes Herbert even dedicated her account to 'the leader of the opposition shoot, soldier, shikari and sometime misogynist'.[67] Although she emphasized the spirit of lively competition, she did not wish to appear too radical, at one point chastising herself for 'talking like a suffragette!'[68] At another moment, after shooting a female wild pig (apparently because it had tusks usually grown by the male) Agnes issued a veiled warning: 'Accidents always do happen when femininity adopts the attributes which are the prerogatives of the masculine gender.'[69]

Despite their playful competition with the all-male party, there is little evidence, however, that the Herberts exercised a radically different attitude to hunting from their male contemporaries. Agnes was both honest and modest about her hunting. Embarking on their first shooting trip, she noted: 'For the first time we said to each other, "let us go out and kill something, or try to".'[70] In much of her writing Agnes was following the literary conventions of figures such as Selous, who had himself pioneered a relatively self-effacing style of adventure writing. Although this style was thus used by both male and female writers, the Herberts gave it a uniquely domestic slant. Thus on one occasion Agnes recounted how she had to be saved from a lion by Cecily's gun and said, just before fainting, 'He will just do for that space in the billiard room.'[71] Agnes also raised questions, in a style unusual in writing by male hunters, about her love of shooting, noting:

> I often wished when I was flushed with success, and I saw my beast lying dead, that I had not done it. It seemed so cruel . . . I began to feel tired of the actual killing as soon as I had perfect specimens of each sort, and always preferred the nobler sport of more dangerous game. I think if I went again I could in most instances deny myself the shot, and content myself with watching and photographing.[72]

Photography was an important part of the Herberts' expedition and, like shooting, was considered as a kind of sport. However, Agnes Herbert's interest in photographing instead of shooting was by no means unique and, as I will show, many others were expressing the idea of using the camera instead of the rifle. Indeed, this was even the case with renowned hunting men like C. H. Stigand who, during an expedition through British East Africa in the same year, took up the challenge to let an approaching rhinoceros get as close as possible to his party before taking photographs instead of shooting the animal.[73] Moreover, although the Herberts did make some photographs of live animals,[74] most show dead animals, skins and 'specimen' heads. What is more, despite the occasional sentiments about cruelty, the Herberts collected altogether trophies of more than eighteen different kinds of game – a bag, they claimed, far more impressive than those of the rival expedition.[75]

As well as their enthusiasm for trophy-collecting the Herberts also shared the racial preconceptions of their male counterparts. Africans were therefore 'savages' whose proper place was in nature. As Agnes asserted: 'The savage who lives in the wild is far more to be admired, and is altogether a more estimable creature than the savage who drives you about Aden, or hauls your boxes about at Berbera.'[76]

While the Herberts were certainly unusual in that they were independent women hunters, in their attitudes towards Africans and their approaches to hunting – with rifles and cameras – they were thus not significantly different from their male counterparts. Although they offered a challenge to the presumption that big-game hunting could only be done by men, they did not suggest any alternatives to the practice.

Photography, Taxidermy and Wildlife

As I have noted, accurate observation, classification and recording had long been regarded as essential techniques for naturalists and hunters alike. During his hunting expeditions in southern Africa in 1835–7 Captain William Cornwallis Harris claimed to have collected not only 'two perfect crania of every species of game quadruped to be found in Southern Africa, together with the skins of lion, quagga, zebra, ostrich &c, tails of the cameleopard and tusks of elephant and hippopotami' but also to have made 'elaborate drawings of every animal that interests the sportsman from the tall giraffe to the minutest antelope'.[77] Later hunters continued this tradition of collecting and picturing. John Guille Millais (1865–1931), son of the artist Sir John Everett Millais, hunted extensively in Africa from 1893 to 1898. Not only did he share his friend Selous's passion for natural history and collecting, but he also put his skills at visual depiction to good use, making detailed paintings and sketches of animals and hunting scenes, many of which were widely reproduced in engravings.[78] As these examples suggest, the idea of realistically capturing, observing and documenting types of animals and human encounters with them preceded the application of photography to the task.

Photography, however, appeared to offer a wholly new, rational basis upon which to represent and classify specimens of flora and fauna, capturing nature in its entirety and in the minutest detail. Hence from the earliest uses of photography on overseas expeditions, animals were targets of the camera. As I have

41 Guy C. Dawnay, 'From the Settite & Royan R.s, N.E. Afr. 1876'.

noted, David Livingstone's Zambezi Expedition (1858–64) used photography as part of a scientific documentation of south-central Africa's resources, which included its fauna. One of the photographs sent by Livingstone to London showed a dead hippopotamus, shot by the explorers, which was clearly intended for Professor Owen's researches into natural history.[79] In the same period, James Chapman took photographic apparatus on his hunting and trading expeditions in the interior of South Africa (1859–63).[80] The photographs Chapman made during his expedition with Thomas Baines to the Zambezi (1860–63) included a number showing animals they shot as well as images of the hunters themselves.[81] These were included in some 140 of Chapman's photographs exhibited at the Paris International Exhibition of 1867, some of which were also sent, via Sir George Grey, to the RGS in the hope of wider recognition of his work.[82]

In the following decades photography became even more widely used to record images of hunting trophies. For men like Guy C. Dawnay, who undertook hunting and exploring expeditions in east and south Africa in the 1870s and 1880s, images of hunting trophies were an important part of personal photograph albums witnessing colonial adventures.[83] A keen hunter, explorer and Staff Intelligence Officer during the Anglo-Zulu War of 1879, Dawnay's love of hunting proved ultimately fatal as he was killed by a wounded buffalo in east Africa in February 1888. Hunters did not only make photographs of themselves with animals they had shot. Dawnay's photograph 'From the Settite & Royan R.s, N.E. Afr. 1876' (illus. 41) shows nine preserved lion heads set up on an easel. The photograph presents a disturbing display of dismembered heads, preserved as if part of live wild animals. The composition, like much contemporary imagery of dead animals, draws in part on the pictorial tradition of *nature morte,* yet it also derives its force from the unique way photographs bring to life, so to speak, objects and moments that can never exist again. In this way, photography was closely associated with the practice of taxidermy: the representation of residues of animals to produce the illusion of live presence.

While taxidermists had been practising their art since at least the eighteenth century, particularly for 'cabinets of curiosity', their number and popularity grew enormously from the 1850s, following increasing interest in natural history and popular fashion. Like photography, taxidermy shared the dual label of art and science. More importantly, both were used as indexical signs to represent past moments of life, producing what Barthes described in photography as 'the living image of a dead thing'.[84] The two practices therefore have parallel technical as well as imaginative associations which would repay closer examination.

Early photographers employed taxidermy in order to capture the wild in a seemingly live pose. As early as the 1850s, for instance, the photographer

J. D. Llewellyn took photographs of stuffed deer, badgers, otters, rabbits and pheasants posed as if caught unawares in the wild. Just as photographers drew on the skill of the taxidermist to overcome their cameras' technical shortcomings, taxidermists drew in turn on the photographer to provide them with an appropriate model of naturalism for their displays. In his popular *Sportsman's Handbook* (1882), the famous taxidermist and publisher of sporting books Rowland Ward (1848–1912) urged sportsmen to take photographic apparatus on their expeditions since 'an animal may be photographed with its surroundings, just as it fell; the picture may be made a nucleus of interesting and most instructive memoranda'. Ward also argued that 'photographic pictures of living *ferae naturae*, in their native jungle or forest ... present the perfect specimen for our contemplation', as they represented the taxidermist's goal of re-creating the lifelike forms of animals within their habitat. To this end Ward recommended the use of the 'Visarap' camera designed by W. G. Tweedy and built by Murray and Heath for sportsmen and travellers to take 'instantaneous photographs of animals and objects in motion'.[85] Ward went on to develop his own 'Naturalists' Camera' for the same purpose.

Rowland Ward's guides on preserving hunting trophies were followed by many hunters and taxidermists who increasingly employed photography as a means of rendering the modelling of animal forms more naturalistic. Indeed, by 1911 Ward could declare:

> the taxidermist could never have reached his present advanced stage without the aid of instantaneous photography ... Previous to the invention of the instantaneous camera I used to have to go to the 'Zoo' and model an animal in wax before I could mount its skin to my satisfaction.[86]

Using photography, Ward developed his own 'artistic setting-up' of trophies and specimens whereby, rather than being 'stuffed', the skin was preserved and remounted on a model constructed from a substance of Ward's own invention, incorporating limb bones and the skull. In achieving a greater attention to detail, expression, gestures and habits, this 'Wardian taxidermy', as it became known, received considerable praise when examples of Ward's work were exhibited at various international and colonial exhibitions, including the Indian and Colonial Exhibition of 1886. One reviewer in *Figaro* declared Ward's 1874 'McCarte Lion', which showed a bullet-wounded lion in a sitting posture, growling with rage and on the look-out for its enemy, to be 'a triumph of modelling' with 'all the artistic powers of a Landseer'.[87] What is particularly interesting here is the way in which Ward used photography – already itself an aestheticization of the 'real' – as the basis for his supposedly naturalistic modelling of animal gesture and expression. Yet as most products of taxidermy at this time showed, from mammals in cases with dried grasses and silk leaves to Rowland Ward's rhino-head drinks cabinet, it was not so much accuracy as illusion that mattered. Just as taxidermists like Ward modelled animals from hunters' photographs, hunters like Dawnay took photographs of remodelled lion heads as a memento of the sight of living wild animals. Indeed, Dawnay's photograph of the lion heads closely resembles the views of Ward's 'McCarte Lion'.[88]

The visual resemblance of Dawnay's photographic composition of animal heads to other photographs of human faces (see, for example, illus. 63) would also not have been lost on contemporary audiences. Indeed, there existed wide associations between the visual representation of humans on the one hand and animals on the other. In his well known romantic and sporting paintings, for example, Landseer often endowed animals with quasi-human expressions. In the field of taxidermy, Herman Ploucquet's displays of stuffed kittens and rabbits wearing clothes and in human-like postures had proved hugely popular at the Great Exhibition of 1851, contributing to an increased demand for such anthropomorphic displays of animals. At a more scientific level, Charles Darwin, for example, used photographs of human facial expressions and bodily gestures in his study *The Expression of the Emotions in Man and Animals* (1872). Here the scientific photographs of facial expression by Dr Guillaume Duchenne and the portraits by Oscar Rejlander were used together with drawings of live animals to represent and compare the physiognomic expression of different emotions, from fear to anger.[89] Dawnay's photograph, and the taxidermist's art it represents, was also an attempt to capture, as if momentarily fixed in time, the natural presence and attitude of live animals.

The association between photography and taxidermy was also derived in part from practices of big-game hunters and naturalists like Frederick Selous who, well versed in using photographs in lectures, frequently also exhibited the skins, horns and stuffed bodies of their actual victims on stage in order to enliven their performances. Indeed, from the 1840s trophy collections, along with illustrated books, were essential in the making of profitable reputations by hunters. Roualeyn Gordon Cumming, for example, author of the highly popular *Five Years of a Hunter's Life* (1850),[90] returned to England in 1849 with some thirty tons of trophies from southern Africa. These were exhibited at the St George's Gallery, Hyde Park Corner, for over two years (1850–52), and many were also shown at the Great Exhibition of 1851. Animal specimens and pictorial displays were integrated in such shows. When Cumming relaunched himself in London as 'The Lion Slayer at Home' in 1855, his entertainment featured not only displays of animal remains but a diorama of thirty African scenes.[91] His hunting paraphernalia was subsequently set up in a popular museum at Fort Augustus in Scotland but was sold, a year before his death in 1866, to the famous showman Barnum for his American Museum.[92] Hunting trophies, along with live zoological specimens and exotic peoples,[93] were thus part of a thriving popular culture of exhibitions in the second half of the nineteenth century.[94] Moreover, many exhibitions, such as the Stanley and Africa Exhibition of 1890, employed a 'trophy' style of exhibiting objects, in which natural products, material artefacts and photographs from different colonial territories were displayed less as objects in themselves than as trophies celebrating the European explorers, administrators and collectors who acquired them.[95]

Similar messages could be conveyed by the resurrection of big game in natural history collections. Hunters like Selous often shot particular animals and preserved their skins in order to see them subsequently resurrected in realistic displays in, for example, the British Museum.[96] In 1907 the big-game hunter and photographer Carl Schillings noted the pleasure to be gained by the European hunter 'when, making a tour of the museums of various places at home, he sees awakened to new life the wild creatures he formerly observed and laid low in far-off lands'.[97] To many hunters, the space of the museum thus replicated the space of the colonial frontier. Natural history museums could even become hunting grounds themselves. During his travels 'with gun and camera' in southern Africa in 1890, Selous's friend H. A. Bryden photographed 'several natural specimens' in the Cape Town Museum, including the head of a white rhinoceros shot by Selous in Mashonaland in 1880.[98] Significantly, nearly all of the photographs of animals reproduced in Bryden's hunting book are of stuffed ones. Stuffed animals had become the ideal photographic target, a re-creation of nature as apparently authentic yet utterly docile.

The animal displays of many natural history museums were also designed to convey imperial messages. The hunter Charles Victor Alexander Peel (*c.*1869–1931), who pursued sport as far afield as Somaliland and the Outer Hebrides,[99] for example, set up his own 'exhibition of big-game trophies and museum of natural history and anthropology' in Oxford in 1906 in order to show 'by means of beautiful objects of Natural History, that there are other countries in the world besides over-populated little England'. Although the museum lasted in this form for only thirteen years before moving to the natural history collection of the Royal Albert Memorial Museum in Exeter, it was a concerted attempt to promote imperial sentiment through hunting and natural history trophies. As Frederick Selous reiterated in his speech to open the museum in July 1906, 'The love of adventure, love of a nomadic life, love of hunting – these instincts had descended to the English-speaking race, and it was due to that that this wonderful Empire had been formed. The hunter had always been a pioneer of Empire.' Peel arranged his museum to provide a tour around hundreds of kinds of animals, collected on his travels, arranged simultaneously as naturalistic displays and as hunting trophies. Arguing that big-game hunting 'exercises all the faculties which go to make a man most manly', Peel hoped that his trophies would inspire young men to go out and 'do good service to the Empire'.[100]

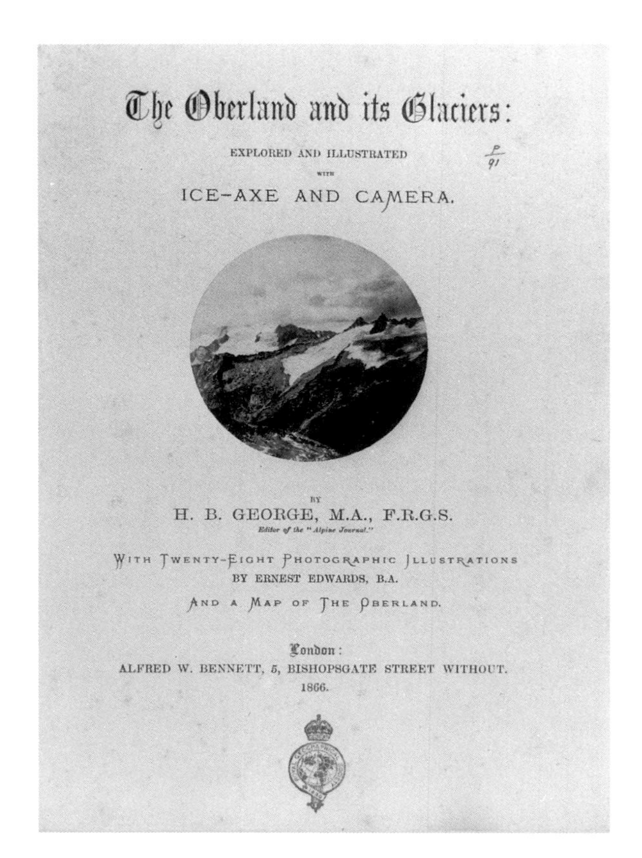
The Oberland and its Glaciers:

EXPLORED AND ILLUSTRATED

WITH

ICE-AXE AND CAMERA.

BY

H. B. GEORGE, M.A., F.R.G.S.

Editor of the "Alpine Journal."

WITH TWENTY-EIGHT PHOTOGRAPHIC ILLUSTRATIONS

BY ERNEST EDWARDS, B.A.

AND A MAP OF THE OBERLAND.

London:

ALFRED W. BENNETT, 5, BISHOPSGATE STREET WITHOUT.

1866.

42 H. B. George, frontispiece and title page, *The Oberland and its Glaciers* (1866).

Similar messages were also conveyed in popular hunting books such as Colonel John Henry Patterson's famous *The Man-Eaters of Tsavo* (1907), in which this English engineer recounted his adventures in 1898 hunting down the man-eating lions which preyed on the Indian and African workers on the Uganda Railway in British East Africa.[101] Interestingly, the frontispiece to Patterson's book is a photograph of lion heads almost identical to Dawnay's 1876 photograph. Patterson's photograph is part of an account which attested to the heroism of the white hunter and his role in maintaining the onward drive of colonial improvement – represented by the Uganda Railway – into British Africa.

By the end of the nineteenth century photographs of 'stuffed' animals, produced for private albums and hunting books, had become almost as common as photographs of dead animals or valuable parts of them. Such photographs were also used in guidebooks for hunters, 'so that he can identify any animal that comes across his path at a glance'.[102] Guidebooks for sportsmen, such as Rowland Ward's *Sportsman's Handbook* (1911), also used photographs of preserved animals to display the 'vital shots' to the hunter.[103]

46 H. J. Mackinder, 'Hausburg on rhino', 1899.

47 H. J. Mackinder, 'Hausburg and gazelle', 1899.

While Mackinder used his Kodak camera on the summit and on the east side of the mountain, Hausburg made a large series of photographs of the whole area.[124] Hausburg's views of Mount Kenya, such as his 'View of Kenya across Valley' (illus. 44), amount to a visual survey of the mountain. Other photographs also correspond with the scientific targets of the expedition, including groups of Masai and Kikuyu people, plants and animals. For instance, Hausburg's photograph 'Alpine Vegetation with Camburn' (illus. 45) shows the expedition's official taxidermist, Claude F. Camburn, who as well as preserving animal specimens, shot animals and identified animals he wanted others to shoot.[125] For as well as photographing and collecting plants, the European members of the expedition shot and photographed numerous animals.[126] Indeed Mackinder linked these two practices in praising Hausburg for being 'so invaluable a shot and so accomplished a photographer'.[127] With the help of porters, the expedition amassed sizeable zoological and botanical collections.[128] Although generally claimed to be for scientific purposes, the hunting of animals was not without an element of sport, as a number of 'trophy'-style photographs suggest. One shows Hausburg sitting, rifle in hand, on a dead rhino he shot (illus. 46) while another shows him, feather in cap, with a dead gazelle (illus. 47). These photographs are less specimens of natural history than trophies of the imperial hunter-explorer.

Mackinder's expedition coincided with British hunting trips in the same vicinity, including those of Admiral Sir Robert Hastings Harris[129] and Edward North Buxton. While busy hunting and photographing lions near Mount Kenya, Buxton caught a long anticipated glimpse of the mountain and, in an eroticized topographic description, noted:

> This was almost the only occasion when this queen of African Mountains unveiled her charms, and I wondered whether Mr. Mackinder, who was then approaching it, would succeed in penetrating that cold barrier which generally covered those ice-filled corries.[130]

Mount Kenya's topography, as much as its flora and fauna, was also in the sights of the scientific weaponry of Mackinder's expedition. Two days before he reached the summit, Mackinder noted in his diary: 'What a beautiful mountain Kenya is, very graceful, not stern, but, as it seems to me, with a cold feminine beauty.'[131] Although he was leading an official expedition for science, rather than a hunting trip for pleasure, like Buxton he was out to capture the topography and animals of Mount Kenya. The 'feminine' nature given to the mountain reinforced the masculine power necessary to 'unveil' its charms, and undertake what he referred to as a 'revelation of its alpine secrets'.[132]

Mackinder was thus following those who, like H. B. George, not only advocated mountaineering and exploring as pursuits essential to imperial manhood

48 'Rock taken from the summit of Mount Kenya by Halford John Mackinder on its first ascent, 13 September 1899'.

but also promoted photography as a tool in the conquest of nature. Reaching the summit of Mount Kenya and capturing it in photographic form represented only the tip of a more detailed and thorough conquest. The making of photographs was accompanied by a cartographic survey of the mountain and surrounding area; the amassing of representative collections of geological, botanical and zoological specimens; the recording of the meteorological characteristics of the region and the collection of observations on the indigenous inhabitants. In short, photographs were not the only trophies of this conquest. Indeed, upon reaching the summit of the mountain, Mackinder took a small lump of granite from the peak. That this rock was less significant as a geological specimen than as a trophy of the conquest of the mountain is evidenced in its subsequent life, mounted on a base to commemorate Mackinder's first ascent (illus. 48). Like other exercises in geographical taxidermy, the photographs taken on Mount Kenya were thus one part of a more complete picturing, collection and appropriation of the natural history trophies of the mountain. Reaching the summit was, after all, the prime goal of this campaign. As Mackinder declared in his journal for 19 September 1899, 'We had snatched victory from what looked likely at one time to be a bad defeat, and now we were sated and impatient for home.'[133]

Mackinder left a scientific imprint on East Africa in the shape of named landscape features, such as Mackinder Valley, and animals.[134] One of the topographical features on Mount Kenya named by Mackinder was Point Thomson, named after John Thomson as a way of acknowledging his considerable photographic services to the expedition.[135] So just as the geography of Mount Kenya was captured with cameras, photography was literally inscribed upon the landscape. These processes of observation, naming and collection represent attempts to incorporate the land, people, flora and fauna into the frameworks of British science.

Camera Hunting

The earliest and easiest means of securing photographs of live wild animals, though perhaps the least intrepid, was at London's Zoological Gardens. A number of commercial photographers followed this route. In 1865 Frank Haes made a series of stereoscopic photographs of animals there.[136] His work was emulated by Frederick York in 1872.[137] Such photography was responding, like the Zoological Gardens themselves, to the popular interest in natural history. As technical developments in photography enabled a reduction in the length of time necessary for an exposure, photographers and naturalists elaborated their experiments with photography to record images of animals.[138] Eadweard J. Muybridge's photographic studies of animals in motion in the late 1870s became particularly well known.[139]

Photographing animals in the wild as opposed to in captivity in zoos or stuffed exhibits in museums was made easier by the technical developments of the 1880s and 1890s: notably the use of new roll film; the increasing portability of cameras; reduction in exposure times and the development of telephotographic lenses. Contemporary advertisements for cameras, telephotographic lenses and telescopes, such as those produced by the London-based firm of Dallmeyer, show the promotion of the use of this technology in practices such as natural history and mountaineering (illus. 49). T. R. Dallmeyer was a pioneer of the telephotographic lens and promoted its use particularly among naturalists, developing his own 'naturalist's camera'.[140] Manuals such as the RGS's *Hints to Travellers*, with their instructions on all kinds of pursuits, including photography and natural history, were ideal publications in which to place such advertisements.[141] Moreover, such improvements in technology were often themselves driven by the contemporary fascination for capturing movement in nature. The French physiologist Etienne Jules Marey (1830–1904) thus turned to photography in the 1880s in his quest to capture movement in nature in graphic form. In 1882 Marey developed a 'photographic gun' to follow seagulls in flight, which photographed as rapidly as 1/720th of a second.[142] Marey was possibly the greatest experimenter in 'time photography' and in the process of his work developed daylight-loading film and the first movie camera with film reels (1886). Unlike Muybridge, with whom he collaborated, Marey was concerned with capturing movement on a single plate from a single point of view. His experiments in the mid-1880s used three cameras to photograph a gull's flight simultaneously from above, the side and in front.[143] That Marey should choose to house his rapid-exposure camera within the body of a gun is particularly interesting given that such photographic advances also coincided with the improvement in hunting rifles and bullets, chiefly through the development of cordite (1893).

49 'Dallmeyer's Portable Telescopes, Photographic Lenses & Cameras for Travellers' in J. Coles, ed., *Hints to Travellers: Scientific and General* (1901).

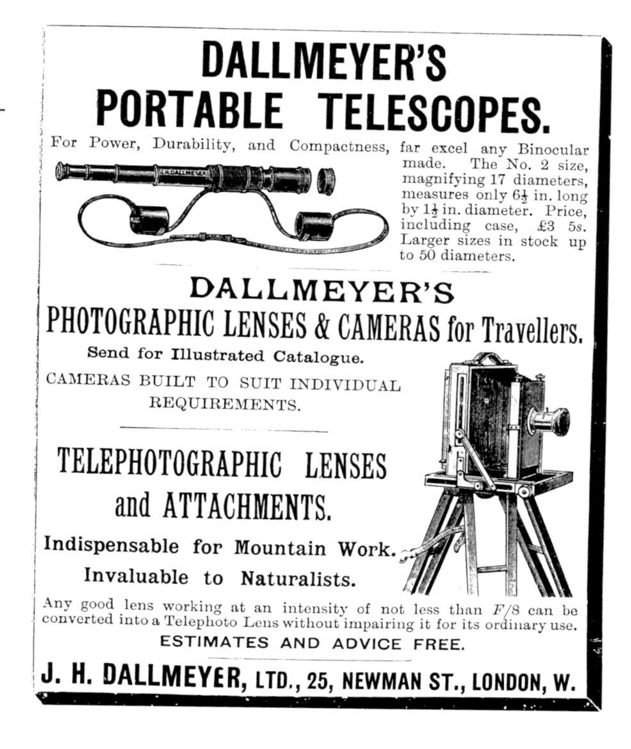

As with Eadweard J. Muybridge's photographic studies of animals in motion, these attempts to capture movement also depended on associations between ballistics and photography. This association has a history almost as long as photography itself, ranging from the experiments of the Englishman Horatio Ross in the 1850s to those of the Frenchman Enjalbert, whose ten-shot spy camera of 1882 closely resembled a revolver and had a trigger-operated shutter. The ever closer parallels between hunting and photography, 'shots' and 'snap shots', were reflected not only in the practices and rhetoric of big-game hunters and photographers[144] but also in the design and advertising of cameras as predatory weapons, some even being designed by big-game hunters.[145]

The use of photography in capturing wild animals was also encouraged by the widespread interest in natural history. Around the turn of the century, popular natural history publications such as *The Living Animals of the World*, with contributions by well-known 'specialists' including F. C. Selous, H. A. Bryden and Sir Harry Johnston, drew on the collections of naturalists and sportsmen to reveal the globe's fauna through countless photographs.[146] Stimulated by such publications and aided by the increased portability of cameras and the development of telephoto lenses, 'nature photography' became increasingly popular, spawning a range of guides and handbooks.[147] Although often aimed at amateur naturalists and photographers in Britain, many such guides drew on the wider popular fascination with 'wild life'. Richard Kearton's popular *Wild Life at Home: How to Study and Photograph It* (1898), for example, contained photographs by his brother Cherry which, with captions such as 'Ascending a thick tree', and 'How to photograph birds' nests in high hedges', showed the intrepid photographer working 'in the wild' at home.[148] Cherry Kearton subsequently embarked on global adventures with his camera, photographing 'wild life', particularly in East Africa.[149] Indeed, it was away from home, in places thought of as truly wild, that photographic hunting came into its own.

Many of the advocates of camera hunting presented it as more sportsmanly than hunting with a rifle. In 1902 the British champion of camera hunting, Edward North Buxton, described it as:

> a field of investigation which promises most interesting results . . . it demands more patience and endurance of heat and other torments, more knowledge of the habits of animals – in a word, better sportsmanship than a mere tube of iron with a trigger; and when a successful picture of wild life is obtained it is a higher achievement, even in the realm of mere sport, than a trophy, however imposing.[150]

Buxton's elevation of camera hunting above ordinary hunting was consistent with his support for conservation movements. A wealthy Englishman with a taste for foreign hunting and climbing adventures, Buxton was a key player in the establishment of the Society for the Preservation of the Fauna of the Empire in 1903. Made up of many well-to-do naturalists and hunters, such as Johnston and Selous, as well as aristocrats and colonial officials such as Curzon, this powerful society was to prove instrumental in generating information on the hunting and trade of animals, the setting up of reserves, and the instigation of game laws and regulations throughout Britain's colonies.[151]

Buxton first used a camera, though without much success, during a hunting trip in the Rockies in the late 1880s.[152] In 1899, he undertook trips to the game districts of British East Africa and the Sudan, 'shooting, photographing and travelling gently from camp to camp'. Partly as a result of these experiences, he became a central figure in the preservation movements being established in Britain in the early twentieth century. On his Sudan trip, Buxton made good use of his new telephotographic 'Dallmeyer's naturalist's camera' (see illus. 49), which he claimed to have used as 'an alternative weapon to the rifle'.[153] However, the camera was still clearly a 'weapon' and only after he had killed all the 'specimens' he required did he devote himself 'wholeheartedly to the absorbing pursuit of camera stalking'.[154]

Nevertheless, Buxton included a good number of his photographs of live animals in his published account. Picturesque photographs of an undisturbed and undefiled natural world, such as 'A Doe Water-Buck' or a group of buck titled 'At the Point of Departure', were placed strategically in the section on game preservation to appeal to the animal-loving sentiments of his readers.[155] Yet his accounts of his excitement at 'camera stalking' serve ultimately to reinforce the continuity between shooting and photographing, with both of them portrayed as adventurous predatory pursuits.[156] While he was concerned with the potential effects of the destruction of wildlife, Buxton was certainly not anti-hunting. Indeed he reassured the 'conservative sportsman' that 'no one proposes to interfere with legitimate sport'.[157]

Buxton received much praise for his advocacy of camera hunting. Harry Johnston proclaimed him to be

> the first sportsman of repute having the courage to stand up before a snobbish public and proclaim that the best sport for a man of cultivated mind is the snapshotting [sic] with the camera (with or without the telephotographic lens), rather than the pumping of lead into elephants, rhinoceroses, antelopes, zebras, and many other harmless, beautiful, or rare beasts and birds.[158]

Although Johnston justified the killing of animals for museums or science, he condemned outright 'the ravages of European and American Sportsmen, which are still one of the greatest blots on our twentieth-century civilization'.[159] In his introduction to C. G. Schillings' *With Flashlight and Rifle* (1906), Johnston claimed that camera hunting, along lines pioneered by Buxton and Schillings, represented 'the sportsmanship of the future'.[160] Carl Schillings, a hunter in German East Africa during the late 1890s, was especially keen to capture in photographs the game he thought would soon disappear for ever. In 1903, having experimented with photographic apparatus (especially magnesium flashlight techniques for night-time photography), he undertook a large-scale expedition making photographs and collecting specimens – live and dead – for German museums and zoos. As well as a retinue of some 170 porters to carry his stores and photographic equipment, he travelled with his own taxidermist (William Orgeich), surgeon and askaris. Using live bait, such as an ox, and trip wires, Schillings secured photographs of lions, leopards and hyenas hunting at night. Using large metal traps placed nearby, he also caught many of the same animals, most of which were shot. By day he operated his camera like a 'photographic gun' following moving animal targets.[161] Schillings described his photographic and specimen-hunting expeditions as attempts to shed 'new light on the tragedy of civilization', in order to help stem the destruction of big game in Africa and to promote conservation along lines drawn up by the British authorities. His work was highly influential within British conservation circles, and had a lasting influence on animal photographers.[162]

Another convert to camera hunting was Arthur Radclyffe Dugmore, an enthusiastic nature photographer[163] who undertook a four-month expedition in British East Africa in 1908 as well as several subsequent expeditions in Africa.[164] Following his first expedition, Dugmore argued passionately for the cause of hunting with the camera, claiming 'unquestionably the excitement is greater, and a comparison of the difficulties makes shooting in most cases appear as a boy's sport'.[165] He claimed that camera hunting was more exciting, skilful and dangerous than shooting, since the camera hunter had to get his cumbersome apparatus within close range of a wild animal while often remaining unprotected. In addition, he said, this 'wonderful sport' was 'clean' and 'wholesome', since it involved 'watching animals and birds in their natural state, living in the country which is their birthright'.[166]

The making of realistic photographs and the collecting of specimens was thus presented as a scientific pursuit for the sake of knowledge and the ultimate preservation of fauna. Many commentators followed Johnston, Buxton and others in asserting the sporting and manly nature of camera hunting. In his preface to Marius Maxwell's *Stalking Big Game with a Camera in Equatorial Africa* (1925), Sir Sidney Harmer, director of the British Natural History Museum, noted that 'camera sport of this kind is no easy occupation, but its devotee must be prepared to risk his life and to stake his quarry in a spirit of adventure not inferior to that shown by the ordinary hunter of Big Game'.[167] Similarly some years later A. L. Butler noted of camera hunting, 'It is far harder to secure such photographs than to shoot the animals, and some of them, taken at close quarters, are evidence of the coolest nerve and finest wood craft.'[168] In the context of new ideas of conservation such claims may be seen as attempts to redefine the codes of sportsmanly conduct.

Since photographs were the trophies of camera hunters, much emphasis was placed on their quality and genuineness. Schillings stressed the accuracy of his photographs, insisting they were 'authentic, first-hand records revealing secrets which the eye of man had never before looked upon . . . the first to show really wild animals in full freedom'.[169] Critical of any attempts to alter or 'retouch' images, Schillings argued that his photographs were 'true to nature' and were absolutely trustworthy 'nature-documents' of African wildlife.[170] Dugmore was equally keen to stress the genuine nature of his photographs and the absence of 'faking' or 'retouching'.[171] These were important facts to be emphasized, given that the value of photographic hunting trophies was contested by some of the keenest hunters. Denis Lyell, for example, was sceptical of the sportsmanliness of photographing game, particularly with a telephoto lens: 'I have no wish to decry the pluck of the telephotographers, as they are brave men, but I do wish to dispel an erroneous idea which is prevalent amongst people who know nothing about hunting.' Lyell claimed in particular that 'charge photographs' (photographs of charging animals) were only 'true' if they were taken from in front of the animal, and that few animals charged unless they were wounded. It was also, he argued, much more dangerous to follow a wounded animal with a rifle than to take photographs of an unwounded one.[172]

In common with other camera hunters, Dugmore suggested that to hunt an animal with a camera meant approaching it at close range unarmed. However, like Buxton and Schillings, he also guaranteed the production of his photographs with a rifle. Indeed, as a photograph of Dugmore with his rifle and camera shows, the two were almost interchangable (illus. 50). Despite his claims to avoid killing animals, while in British East Africa Dugmore put himself in positions to get dramatic photographs that more than occasionally resulted in the animal having to be shot. One photograph showed a rhinoceros apparently 'photographed at a distance of fifteen yards when actually charging the author and his companions' (illus. 51). However, as Dugmore added, 'as soon as the exposure was made a well placed shot turned the charging beast'. He described hunting this rhinoceros thus: 'When he seemed as close as it was wise to let him come I pressed the button, and my companion, as agreed, fired as he heard the shutter drop.'[173] This was by no means an unusual practice for camera hunters; Carl Schillings had similar encounters with rhinoceros where cameras and guns were shot simultaneously.[174] Dugmore's stance with gun and camera,[175] like the photograph of the camera hunter Marius Maxwell with his camera and dead elephant (illus. 52), highlights the ambiguous place of these camera hunters. In their very attempt to photograph wild and living nature, they were capturing and re-creating in photographs the experience of hunting and killing.

Inspired by Schillings and Dugmore, Maxwell himself set out for East Africa in 1911 and again in 1921, to 'secure photographic records of incidents in big-game hunting . . . to obtain an accurate shot with the camera instead of the rifle'.[176] Although Maxwell claimed to use the rifle only in extreme cases, his attempt to photograph elephants in the wild 'exactly as the hunter would see it' led him to shoot elephants with both the camera and rifle.[177] His 'A Victim' is the final photograph in a sequence which re-creates the approach of a herd of elephants and the shooting of the biggest bull.[178] This image, showing Maxwell leaning against a dead elephant with a camera under his arm, encapsulates in striking terms the metamorphosis of the camera into the gun and the practice of photography as a predatory pursuit.

The contradictions implicit in much camera hunting were highlighted in the early 1920s by the wildlife photographer Cherry Kearton in his book *Photographing Wild Life Across the World*. Kearton abhorred not only the 'wanton destruction of wild animals' by big-game hunters but also the activities of the photographic hunters who were 'guilty of maiming their "sitters" by gun-shot'.[179] Although he did not name his culprits, both Schillings and

50 Arthur Radclyffe Dugmore, 'The Author and his camera', *Camera Adventures in the African Wilds* (1910).

51 Arthur Radclyffe Dugmore, 'Rhinoceros photographed at a distance of fifteen yards when actually charging the author and his companions', *Camera Adventures in the African Wilds* (1910).

52 M. Maxwell, 'A Victim', *Stalking Big Game with the Camera* (1925).

Dugmore made photographs of wounded animals.[180] At issue, according to Kearton, was the violence of

> a certain type of photographic expedition or safari, which, while pretending to forward the interests of Natural History, frequently takes as big a toll of animal life as the Big Game hunter proper, who goes out with the sole and frank idea of collecting specimens.[181]

The charges of a generation of camera hunters were now turned on themselves.

The contrast between activities of individual camera hunters in the late nineteenth and early twentieth centuries and the large-scale, professionalized photographic safaris in the 1920s and 1930s in British colonial Africa is sometimes interpreted as signalling a wider transformation in European attitudes towards wild animals – a shift away from indulgent slaughter to enlightened conservation. However, photographic hunting was far from exempt from cruelty or killing. Furthermore, as photographic safaris became even more bulky than simple shooting trips, they still commanded the labour of Africans as trackers, guides and porters. Although cameras replaced guns, 'gun bearers' simply became 'camera bearers'.[182] Similarly, gun laws, game laws, administrative policies and game reserves had transformed indigenous African hunters into 'poachers'. Organized photographic hunting in fact marks a shift in the terms of domination, away from a celebration of brute force over the natural world to a more subtle though no less powerful mastery of nature through colonial management and stewardship. This shift is inescapably linked to the broader colonial transformation of Africa itself, from an era of exploration and conquest to one of settlement and administration.

East Africa, which big-game hunters and naturalists regarded as a hunter's 'paradise' and 'Garden of Eden',[183] was the quintessential domain for the development of camera hunting. Opened up by colonial rule and the Uganda Railway, British East Africa became the prime hunting ground for wealthy tourists and collectors often working on behalf of natural history museums in Europe and America.[184] Although they were its chief architects, the British were by no means the only camera hunters. The American taxidermist and naturalist Carl Akeley used the concept and technique extensively during his visits to East Africa (in 1905, 1909 and 1921) for the American Museum of Natural History. By collecting photographs and animal skins he reconstructed animals for the museum, styling himself as a 'sculptor and biographer of the vanishing wildlife of Africa'. Like British preservationists, Akeley particularly commended hunting with cameras, since 'when that game is over the animals are still alive to play another day. Moreover, according to any true conception of sport . . . camera hunting takes twice the man that gun hunting takes.'[185] The fact that camera hunting was common within the work of both European and American naturalists shows that photographic technology and the language of hunting were not merely part of British colonial discourse, but were implicated in broader movements to create and preserve a vision of nature as a timeless domain for white Euro-American men.[186]

Nevertheless, the conservation discourse of which camera hunting was part was also implicated in the politics of colonial rule in East Africa, which excluded Africans from their traditional hunting grounds, controlling them through new spaces of 'reserves'. Hunting, whether with rifle or camera, was the preserve of the white man. Denis Lyell explained this in terms of men's differing aptitude for 'deductive reasoning', adding, 'that is why the native is often so inferior to the white man'. Without powers of deductive reasoning, Lyell argued, local knowledge of nature could not aspire to scientific observation.[187] For men like Stigand and Lyell, Africans could not be 'sportsmen' since they hunted only for love of meat and for the lust of killing.[188] Thus, although useful for tracking animals, Africans had no 'sporting instinct' comparable to that of Europeans.[189]

These attitudes helped maintain the imaginary boundaries between white and black, civilized and uncivilized, human and animal. For many British men like Stigand, the entire African empire could be conceived as a project of animal management. In 1902 Buxton, champion of big-game preservation and camera hunting, had noted, 'We are establishing Reserves in which all kinds of wild beasts are to be left to fight it out. Can we not extend such a measure to some of the human species, to this extent – that they shall govern themselves, and the strong shall prevail?'[190]

An underlying faith in scientific theories of racial evolution and the 'struggle for existence' allowed the characterization of Africans as closer to the animal ancestry of the human race than Europeans. Hunters like Stigand were even more direct in their equation of Africans with animals. While he argued that 'native policy' should be tailored to suit the specifics of particular areas and tribes, his general view was that 'many natives must be treated more as animals than human beings – treated with kindness, but no attempt made to put them on the same plane as ourselves'.[191] Indeed, he advocated an extreme from of colonial management and eugenics in which the white man placed the 'best tribes' (those that were 'better disciplined, more hard working, better organized and more intelligent') in large reserves in order to slowly civilise themselves under the supervision of the white man. Meanwhile the 'indolent and worthless' tribes would be forced to supply labour for this protectorate, 'especially such parts as were decided on for white settlement and plantations'.[192]

Hunting with the camera similarly continued to coexist with a colonial way of seeing. This is clear, for example, from the hunting, pioneering and photographic activities of men such as Major William Foran (1882–1966) who worked in British East Africa in the first two decades of the twentieth century as a police officer and soldier. A keen amateur photographer, Foran was a close friend of Stigand, having attended the same 'Army Crammers' and served with him in East Africa.[193] Furthermore, Foran shared much of Stigand's colonial vision. In his book *Kill or be Killed: The Rambling Reminiscences of an Amateur Hunter* (1933) Foran recalled his times hunting and pioneering in British East Africa in the early part of this century. As the title suggests, he regarded hunting as an expression of a natural warfare. He began his book, 'It is thirty-four years since I fired my first shot in Africa at either man or beast, and thus found myself launched in the everlasting warfare waged between all living creatures.' Thus, like Stigand, Foran regarded his hunting and campaigning as mutually supportive and ultimately part of the same process of colonial ordering. Foran also found photography a useful weapon in this campaign. As he wrote in 1933: 'for the first six years of my hunting, I employed mostly a rifle to provide me with sport. Thereafter, I relied almost solely upon a camera.'[194] His preference was not the result of any pleading for photography as a form of conservation but quite simply because it offered more sport. In his own words, 'Adventure means much to all red-blooded he-men, but too often it is come by far too tamely and easily . . . novelty was introduced by Carl. G. Schillings, A. Radclyffe Dugmore and Cherry Kearton in the substitution of the camera for the rifle.'[195]

Although Foran adopted the camera, his hunting remained a deadly pursuit of his prey. Like many camera hunters he wielded rifle and camera simultaneously; animals were often shot with the camera only moments before they were shot with the rifle.[196] This is not such a paradox if one remembers that early preservationists were rarely anti-hunting but simply advocates of restricted access of game to proper 'sportsmen'.[197] Camera hunting fitted neatly into this new code of hunting sportsmanship, as its supporters had consistently emphasized its skill and manliness. Indigenous people, needless to say, were excluded from this new preservationist hunting code. Indeed, for many hunters, soldiers and pioneers, 'savage' peoples were understood as a form of wildlife to be conquered and controlled. The representation of colonized people as somehow bound within the natural world served to distance the space of the 'wild' and 'savage' from that of the 'civilized' and colonized. These boundaries marked conceptions of cultural difference which, as I go on to explore in the next chapter, were central to the ways in which photography was used to capture and categorize racial 'types'.

5 'Photographing the Natives'

Visitors to the first exhibition of the Photographic Society in London in 1854 were, according to one reviewer in the *Art Journal*, greatly impressed by a range of human portraits, including 'the Zulu Kaffirs' by a Mr Henneman and 'the insane' by Dr Hugh Diamond.[1] The ability of photography to apprehend visually the human body made it one of the most powerful mediums in the Victorian era for bringing the British viewing public imaginatively face to face with people different from themselves, both abroad and at home. Commercial photographers in particular tapped into a market hungry for portraits of exotic and little-known peoples from around the world. Much commercial activity, although often focused on the picturesque, was also closely associated with discourses on ethnology and anthropology. Such pursuits owed their existence in large measure to colonial expansion and the resulting exposure of different races to European eyes.[2] The institutionalization of anthropological inquiry through the establishment of organizations such as the Aborigines Protection Society (1837) and the Ethnological Society of London (ESL) (1843) was accompanied by an increasing concern with securing accurate and reliable anthropological information.

Much of this enterprise, and the place of photography within it, was motivated by the belief that Aboriginal races were vanishing before the onslaught of 'civilization' and such peoples and their culture ought to be recorded urgently before they disappeared for ever.[3] Evidence of this 'salvage motive' shaping photographic practice is perhaps best given in the photographs made of the remnants of the inhabitants of Tasmania, whose indigenous population had declined drastically, due largely to new kinds of disease, since British settlement at the beginning of the nineteenth century. Francis Nixon, the Bishop of Tasmania, photographed the remaining Aborigines in 1858, when they numbered only fourteen or fifteen, in the settlement at Oyster Cove, near Hobart, where they were imprisoned. It was possibly the display of Nixon's photographs at the International Exhibition in London in 1862 that led to the professional photographer Charles Woolley being commissioned to photograph the few remaining Tasmanians for display at Tasmania's stand at the Intercolonial Exhibition in Melboune in 1866. Woolley's photograph of

53 C. A. Woolley, 'Trucanini, 1866'.

Trucanini (illus. 53) was one of five chosen for the exhibition from a set of fifteen photographs of five Tasmanian Aborigines made by Woolley in his Hobart studio in 1866. Although there was some dispute over how representative Woolley's portrait was of the Tasmanian Aborigines in general and Trucanini in particular, this portrait was widely distributed in photographs and engravings in books such as James Bonwick's *Last of the Tasmanians* (1870). Through such displays the picture came to stand as an icon of a vanished race, particularly when Trucanini herself became the sole remaining survivor in 1869, after which she lived for only a further seven years.[4] The photograph's appeal as a melancholy view of the 'last of the Tasmanians' is en-

54 Anon, 'Photographing the natives', *Photographic views of Blantyre*, c. 1901–5.

hanced both by Trucanini's haunting stare and by the use of the vignette, a popular Victorian convention, to isolate and dissolve the portrait into the surrounding emptiness.

Although the photograph of Trucanini is a portrait of a named individual and moreover one whose defiant stare may also disarm attempts to classify and objectify her as a specimen of 'the race',[5] the photographing of disappearing races was an important part of a broader exercise in objectification and survey. Indeed, this portrait was one of three standard poses (full-face, profile and three-quarter-face) made by Woolley of each sitter, as requested by the commission in line with contemporary demands for a regularized system of photographing racial 'types'. The objectification which was entailed in such anthropological inquiry is also apparent in the uses of other traces of Trucanini, not least her own body. Indeed, Trucanini's body was exhumed two years after her death and, despite agreement that her bones were not to be shown in public, her skeleton was set up as an exhibit in the Tasmanian Museum in 1905.

By the turn of the century 'photographing the natives' was an activity common enough to itself be the subject of photography (illus. 54). The making of such photographs constituted an important part of the colonial encounter. For explorers and missionaries to Africa in particular, photographic apparatus was seen as a sign of European cultural superiority with which to impress and even intimidate indigenous peoples. Optical devices in general, from looking-glasses to cameras, were influential within the European encounter with Africa at least in part because of the significance of metaphors of light, sight and vision with Western Christian conceptions of 'self'.[6] The explorer Joseph Thomson used his photographic apparatus with 'astonishing results' on his East Central African Expedition of 1878–80, along with his usual tricks revealing and concealing his artificial tooth. Thomson found the local people so frightened of the camera that 'by leaving a camera standing alone he had kept a whole village totally deserted for a day'.[7] Thomson made further attempts to photograph 'the natives' for purposes of ethnology while leading an expedition to Mount Kenya and Lake Victoria Nyanza in 1883–4:[8]

> With soothing words, aided by sundry pinches and chuckings under the chin, I might get the length of making them stand up; but the moment that the attempt to focus them took place they would fly in terror to the shelter of the woods. To show them photos, and try to explain what I wanted, only made them worse. They imagined I was a magician trying to take possession of their souls which once accomplished they would be entirely at my mercy. They would not in the end even look at a photo . . . I spoiled several negatives, and finally gave up the attempt.[9]

Late-nineteenth-century photographers similarly found the Tswana of southern Africa reluctant subjects because of the belief that the photograph separated a person from his image and hence could capture the 'self'.[10]

It was not just in Africa that travellers encountered such fears of the camera. During his travels in China in the 1860s, the commercial photographer John Thomson (no relation to Joseph) reported that many Chinese believed that the camera could see through the landscape, and that the photographic process involved the use of eyes stolen from children.[11] General suspicions of the camera and its operator were not perhaps entirely unfounded, since those sitters whom photographers did manage to capture had neither knowledge nor control over the uses and meanings of their likenesses. Thus to a large extent explorers' accounts of superstitious fears of the camera were themselves misreadings of cultural difference and the very real threat that their presence and technology posed. Perhaps the more important question then lies less with 'native' reactions to the camera than, as Michael Taussig neatly puts it, 'with the white man's fascination with their fascination with these mimetically capacious machines'.[12] To a large extent this fascination was derived from the power it gave those who wielded such technical devices. Both commercial photographers such as John Thomson and explorers like Joseph Thomson exploited the effect their photographic technology had on the people they encountered. John Thomson thus noted in the introduction to his *Illustrations of China and Its People*:

I therefore frequently enjoyed the reputation of being a dangerous geomancer, and my camera was held to be a dark mysterious instrument, which, combined with my naturally, or supernaturally, intensified eyesight gave me power to see through rocks and mountains, to pierce the very souls of the natives, and to produce miraculous pictures by some black art, which as the same time bereft the individual depicted of so much of the principle of life as to render his death a certainty within a very short period of years.

It is perhaps not surprising, then, that photographers attempting to secure pictures of indigenous people encountered no little resistance. Indeed, during his travels in China John Thomson was attacked with stones and roughly handled on more than one occasion. As he later recalled of his travels in general, he 'frequently found that natives of foreign countries resent the liberty taken of pointing a camera at them, and fly as if they expected to be shot'. To avoid such difficulties he recommended a hand-held camera which 'may be used when the operator is facing at right angles to the object to be photographed'.[13]

This concern with securing photographs without the knowledge of the subjects being photographed is perhaps an appropriate metaphor for the ambition of regimes of colonial representation: to see without being seen. As Timothy Mitchell has suggested, in the context of European colonialism in the Middle East, 'the photographer, invisible beneath his black cloth as he eyed the world through his camera's gaze ... typified the kind of presence desired by the European whether as tourist, writer or indeed colonial power'.[14] Although Victorian photographers, with their complex technical manoeuvrings and cumbersome equipment, were themselves highly conspicuous figures, this kind of one-way vision was often re-created in photographs for the viewer to enjoy.

This kind of presence was also enacted through other kinds of scientific technology. Francis Galton exemplified this during his expedition in Africa in 1852–4 when he exercised his passion for exacting measurement on an African woman. In his account of his expedition, Galton referred to a Nama woman, the wife of one of his host's servants, as a 'Venus among Hottentots'. Being 'perfectly aghast at her development' and being a 'scientific man', Galton was 'exceedingly anxious to obtain accurate measurements of her shape'. However, the circumstances were difficult and Galton reported that he 'felt in a dilemma as I gazed at her form'. The solution, of which Galton was particularly proud, involved his taking a series of observations 'upon her figure' with his sextant, making an outline drawing while she stood at a distance under a tree. Then he 'boldly pulled out' his measuring tape to calculate the distance and worked out the results using trigonometry and logarithms.[15]

Galton's fascination with the body of a black woman as an object of desire and his employment of European scientific instruments to comprehend its difference from a distance are significant for a number of reasons. For a start, they

embossed ruler on their covers. It was in such a climate that some notable attempts were made to link directly photography and anthropometry, which by 1840 had come to mean the measurement of the living human body with a view to determining its average dimensions at different ages and in different races or classes.[39] In 1869 Jones H. Lamprey, Assistant-Secretary of the ESL and Librarian of the RGS, devised a measuring screen for ethnological photography consisting of a grid of string divided into regular two-inch squares, against which the naked human subject was photographed in full-length, front and profile views (illus. 55).[40] This scheme was partly an application of a well-established pictorial device to aid the accurate delineation of the human figure.[41] However, Lamprey's method was intended to provide a framework for categorizing and comparing distinct 'racial types'. Through such photographs, he argued, 'the anatomical structure of a good academy figure or model of six feet in height can be compared with a Malay of four feet eight in height.[42]

The screen isolated the human subject from a wider cultural environment, placing the body within a regularized and measurable lattice – a 'normalizing grid'.[43] The cartographic rhetoric and the flatness of the image construct it as scientific and objective. Lamprey suggested that setting the human subject against these 'longitudinal and latitudinal lines' would enable the study and definition of 'all those peculiarities of contour which are so distinctly observable, in each group'.[44]

This form of mapping was part of a process of anthropometry through which cultural difference was made visible in the human body and rendered knowable. Lamprey was particularly interested in the ethnology of the Chinese, having travelled in China in the early 1860s when it was increasingly coming under the gaze of British imperial interests.[45] Yet Lamprey intended his method to be part of a much more extensive project to map different races. He argued that 'photographers on foreign stations would greatly assist us if they adopted the same plan' and could himself claim 'my portfolio already contains a collection of specimens of various races'.[46] Examples included photographs of men from China, West Africa, Madagascar and Poland.[47] Photographs made by Lamprey's method also featured in Carl and Frederick Dammann's *Ethnological Photographic Gallery of the Various Races of Man* (1876), a massive work containing fifty plates with grids of ten to fifteen photographs on each plate.[48]

Lamprey was not alone in formulating schemes of anthropometric photography designed for Empire-wide use. In the same year Thomas Henry Huxley, President of the ESL, proposed a scheme for producing a 'systematic series of photographs of the various races of men comprehended within the British Empire' to the Colonial Office.[49] His method consisted of making front and profile photographs of the human subject without clothes. Portraits were to be taken at a standard distance from the camera and subjects were to be

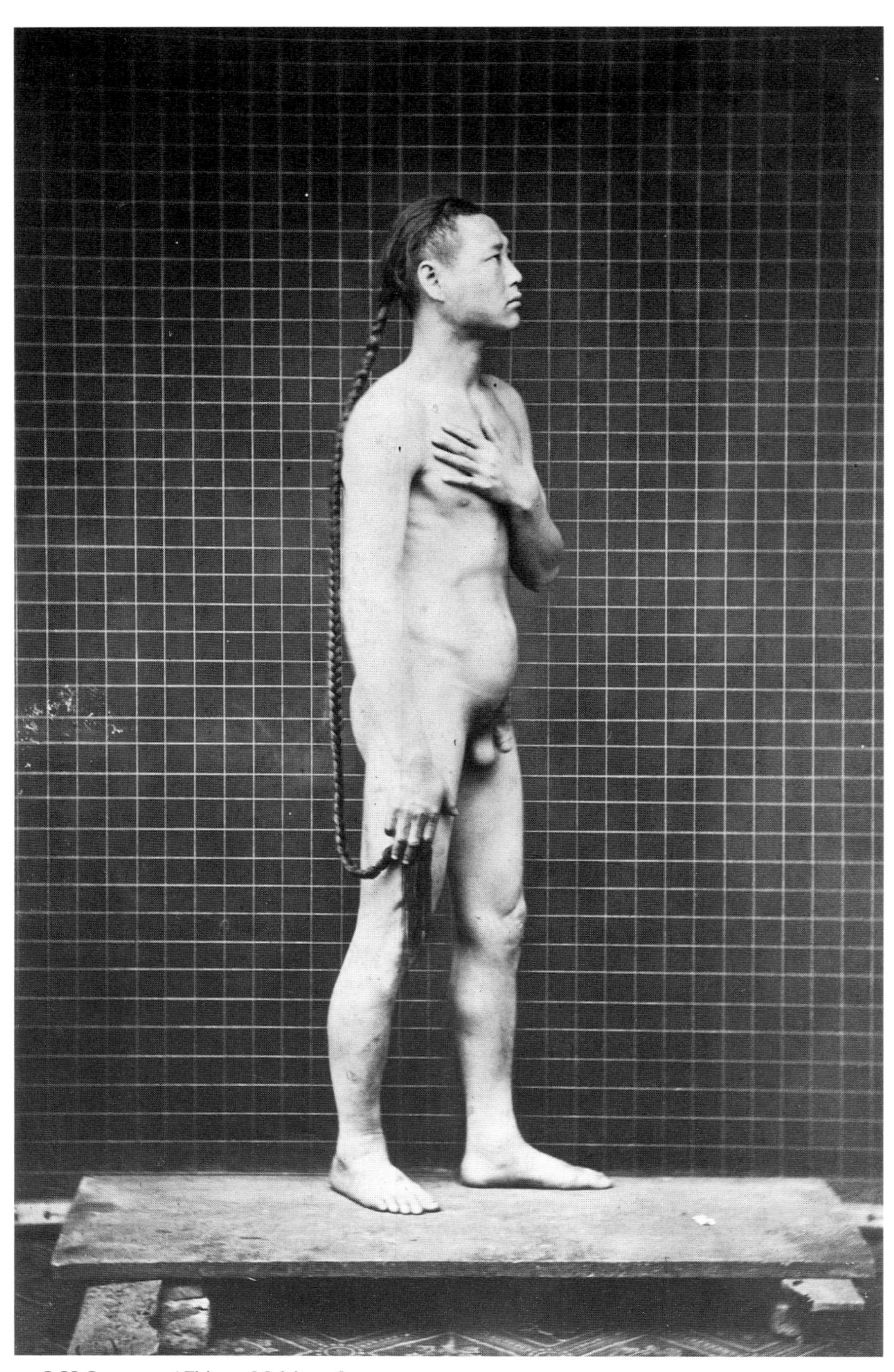

55 J. H. Lamprey, 'Chinese Male', *c.* 1870.

photographed in standard anthropometric poses with a clearly marked measuring rod.

While the Colonial Office sent out a circular in 1869 appealing for anthropological photographs of the indigenous inhabitants of the colonies, no systematic programme of the order Huxley had imagined materialized. To subject people to such rigid demands required the imposition of fairly precise forms of bodily control and it is significant that much of the photographic work executed precisely along Huxley's lines was limited to strictly controlled groups, such as convicts in the Breakwater Prison in South Africa.[50] Not only was it difficult for photographers and anthropologists to secure the co-operation of colonized peoples for Huxley's scheme, but the scheme also failed to provide reliable data on more complex anatomical measurements – for example, of 'stature'.[51] Huxley's method was not as widely adopted as that of Lamprey by photographers in the field. While systematic attempts to apply photography to anthropometry were not perhaps made on the vast scale envisaged by their formulators, they further established the idea of the racial 'type' and the usefulness of the camera in capturing it.

The photographic techniques and survey ambitions of Lamprey and Huxley certainly influenced the anthropological work of men like Maurice Vidal Portman (1861–1935), Extra Assistant Superintendent at the penal settlement of Port Blair on the Andaman Islands from 1879 to 1899. The Andaman Islands, a chain of islands in the Bay of Bengal, began to be incorporated as a British colonial territory following the establishement of the penal colony at Port Blair in the 1850s. As 'Officer in Charge of the Aborigines' Portman was involved in attempts to extend the British colonial and civilizing influence among the indigenous inhabitants. This involved not only making contact with islanders, by capture if necessary, but also the running of 'Andaman Homes', initially established near Port Blair in 1863, designed to encourage the Andamanese to adopt a settled, economically productive way of life. Much of Portman's communication with the islanders was therefore undertaken as a way of ensuring their co-operation with British geographical exploration, survey and settlement.[52] As well as aiding the topographical survey of the islands, Portman undertook a comprehensive study of the Andamanese for the British Museum in the late 1880s and early 1890s. This ambitious survey, undertaken with the assistance of W. Molesworth, a surgeon in the Indian Medical Service, resulted in eleven volumes of detailed notes, photographs and statistical information.[53]

As I have noted, much anthropological work was fostered by a concern with recording indigenous races before they died out. Portman's survey was partly motivated by the belief that the Andamanese were on the inevitable path to extinction. Consequently men like Professor William H. Flower, Director of the Natural History Museum, argued that the destruction of 'primitive races'

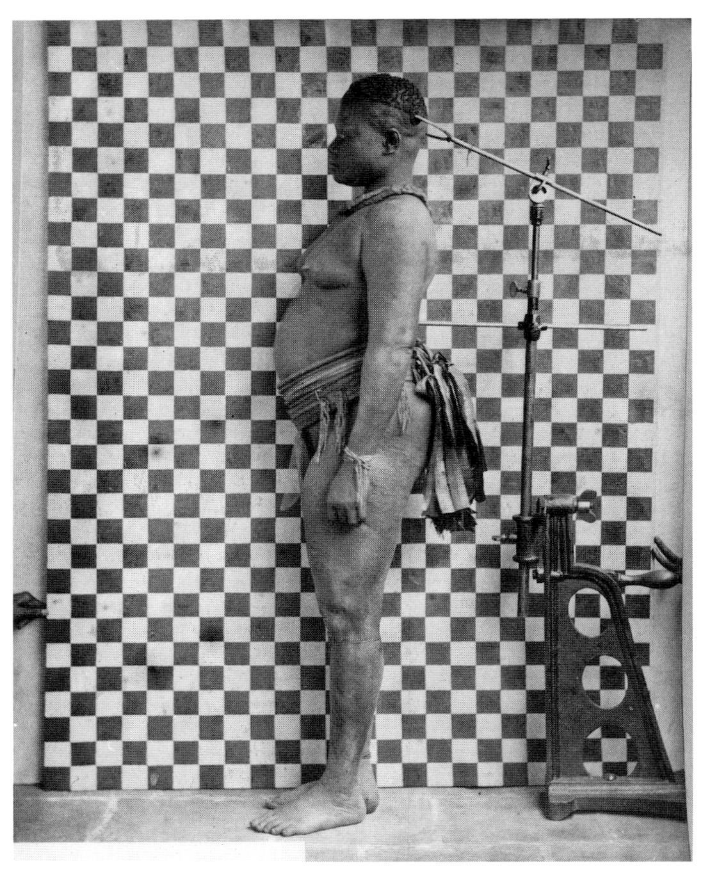

OBSERVATIONS ON EXTERNAL CHARACTERS.

Preliminary Particulars.

No. 12 Date 28th June 1893.

Sex Female. Age 42 (about).

Tribe Puchikwár (Éremtága). Locality Yéretil

Language or dialect Puchikwár. Interior of the middle Andaman.

General condition 2

(1) Stout. (2) Medium. (3) Thin

Descriptive Characters.

(A.) Colour of skin on parts not exposed to the air? ... ... 3 — 29 in Broca's scale.

Black { Absolute (1.) Sooty (2.) }
Brown { Reddish (3.) Yellowish (4.) }
Red (5.)
Yellow (6.)
White { Yellowish (7.) Brownish (8.) Pale (9.) Rosy (10.) }

(For colour of skin *see* plate II., Figs. 1—10.)

(B.) Colour of eyes? ... ... Very dark brown.
Dark of all shades (1.)
Medium { All medium shades, expect green (2.) Green (3.) }
Light { All light shades except blue (4.) Blue (5.) }

(For medium colours of eyes *see* Plate II., Figs. A, B, C, and D.)

(C.) Fold of skin at inner angle of eye? 3
(0) Absent. (1) Vesting remaining. (2) Covering $\frac{1}{8}$ to $\frac{1}{4}$ the caruncle. (3) Covering the caruncle

(D.) Colour of hair? 1
(1) Black. (2) Dark brown. (3) Medium. (4) Blond or Fair of all shades. (5) Red.

(For medium shades of hair *see* Plate II., Figs. A, B, and C); if possible a lock of hair should be attached to the Schedule.

(E.) Character of hair? 4
(1) Straight. (2) Undulating or way. (3) Curly. (4) Woolly

12.

56 Maurice Vidal Portman, 'Burko. Profile View of the Same Woman' ['Woman of the Ta-Keda tribe, age about 40 years'], *c.* 1893.

57 Maurice Vidal Portman, 'Observations on External Characters: No. 12' ['Woicha, Woman aged about 42'], *c.* 1893.

through absorption and extermination made the collection of anthropological data through 'an institution commanding the resources of the nation' an urgent task.[54]

The Andamanese were regarded as particularly important since their geographical isolation and low status on the evolutionary ladder made them representative of the 'childhood of mankind'. In 1880, for instance, Professor Flower described the Andamanese race as 'an infantile, undeveloped or primitive form of the type from which the African Negroes . . . and the Melanesians . . . may have sprung'.[55] Flower thus offered particular encouragement to Portman's researches on the Andamanese, urging him to 'lift up the veil behind which they had been so long concealed'.[56]

Portman's survey of the Andamanese was certainly intended to be comprehensive. His use of anthropometric photography (illus. 56) suggests that he adopted a method similar to that advocated by Lamprey in 1869. Portman's canvas screen, with painted checks each two inches square, like Lamprey's string grid, placed the subject within a cartographic lattice in order to generate reliable, comparable scientific data. Portman's anthropometric photographs were accompanied by two volumes of 'Observations on External Characters' which provided detailed information on the physical characteristics of those photographed, from weight and skin colour to ear length and pulse rate (illus. 57).[57] In addition, tracings of a hand and foot of each individual were made. Together, this mass of intimate measurements, tracings and photographs was designed to capture and preserve the race as completely as possible. Individuals were also categorized by their age, sex, tribe, language and locality. The latter related to Portman's general geographical classification of the Andamanese into the North Andaman, South Andaman and Öngé groups of tribes.

Portman's anthropometric photographs and measurements were complemented by information on the temperament and disposition of those pictured. Thus 'special observations' for 'Woicha' included 'very quiet and pleasant disposition. Fairly intelligent'. More significantly, the photographs themselves, like the traces of hands and feet, were represented as direct measurements of the intelligence and temperament of an individual body. As Portman noted, 'Especial intelligence . . . is usually accompanied by refined good features, particularly nose and mouth, also by an irritable temper indicating a nervous temperament.'[58] Such guidance applied particularly to Portman's series of front and profile portraits of the faces and heads of selected Andamanese, made in addition to the anthropometric measurements.[59] These portraits are accompanied by labels which give the name, tribe, locality and disposition of individuals, noting, for example, 'Woichela. Man of the "Áka-Juwai" inland tribe, Middle Andaman. Age about 31 years. He is quiet, and of good temper: very plucky, and docile.'[60]

Portman's guiding framework for constructing the photographs and their descriptive captions was, as he himself noted, provided by *Notes and Queries*.[61] In this he was following others who had made anthropological photographs of the Andamans, particularly E. H. Man.[62] Portman argued that photographs, accompanied by explanatory texts, offered the most satisfactory means of providing a scientific record of comparable data as set out in *Notes and Queries*.[63] As well as providing detailed information on the physical character of 'types', Portman tried to capture in photographs the development of Andamanese women over time. Thus an individual anthropometric photograph of an Andamanese woman (see illus. 56) was part of a series of full-length, front and profile photographs of women, beginning with a girl of six years old and ending with a woman of about sixty-five.[64] With such a racial panorama, Portman showed that photographs of racial 'types' at different ages could stand for the development of a single racial 'type'.

As well as making anthropometric photographs and photographs of individual faces, Portman also used photography to document various kinds of Andamanese manufacturing techniques. His survey includes sequences of images showing the making of an adze (a small axe), as well as bows, arrows, ropes and huts.[65] Although such images do not construct a 'type' through measurement of physical difference, many of them tend to remove individuals from a wider lived environment and are used as evidence of typical features of their general character.[66]

Other images present a more contradictory picture. A heavily retouched photograph of 'Group of St [South] & Middle Andamanese' (illus. 58) shows Portman surrounded by a group of Andamanese. In a scene reminiscent of photographs of theatrical troupes, Portman is certainly the central player. This would be consistent with his representation of himself to the RGS – to whom the photograph was presented in 1888 – as a friend of the Andamanese: 'Their association with outsiders has brought them nothing but harm, and it is a matter of great regret to me that such a pleasant race are so rapidly becoming extinct.'[67] Countering claims that the Andamanese were cannibals, Portman represented them as simple, harmless and fun-loving: 'Being so low in the scale of civilization, none of the vices of civilization are known to them.'[68] Despite this protective stance, rather than adopting a commanding or paternalist position overseeing the group, as in the photographs of E. H. Man,[69] here Portman is reclining on the ground. In contrast to his anthropometric photographs this image, with its overlay of paint and text, presence of the photographer and lack of measuring screen, appears unscientific. Yet it might operate within a developing alternative set of ideas of anthropological accuracy. In his 1896 article on photography for anthropologists Portman drew attention to the advice in *Notes and Queries* that 'savages will be found to answer more freely when the interrogator places himself on the same level as themselves, i.e. if they sit upon the ground he should do the same'.[70]

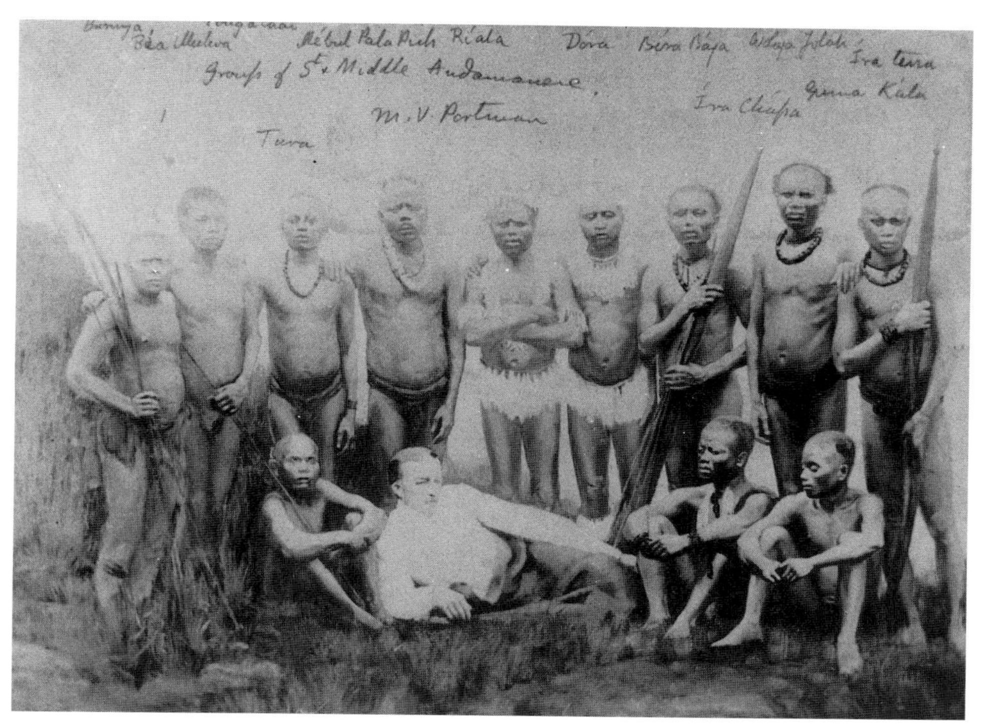

58 Maurice Vidal Portman, 'Group of St. & Middle Andamanese', 1887.

Portman's pose in this photograph and his awareness of the symbolism of such spatial relations clearly place him at least on the same level as the Andamanese. Despite his view of them as uncivilized, his naming of individuals on the image itself suggests a personal commitment beyond the interests of scientific classification; these figures are not 'types', but individuals.

The People of India

One of the most ambitious photographic surveys of racial types, which also exemplifies the complex links between anthropological photography and colonial administration, was *The People of India*. This massive work, in eight volumes with 468 photographs in each set, began as a private collection of Lord Canning, Governor-General from 1856–8, but was transformed after the Indian rebellion of 1857 into an official project of the Political and Secret Department of the India Office. Indeed, of the 200 sets published between 1868 and 1875 half were retained for official use.

Edited by John William Kaye and J. Forbes Watson of the India Office, *The People of India* was intended 'fairly to represent the different varieties of the

Indian Races'.[71] Yet the photographs for *The People of India* were not collected in any systematic form. Rather, they were amassed from the wide-ranging portrait work of a number of photographers, including amateur enthusiasts such as Willoughby Wallace Hooper[72] and commercial photographers such as Charles Shepherd and James Robertson. Photographs originally made for a variety of purposes were thus put together in a large-scale descriptive survey. The editors were aware of the lack of consistency, noting, 'although the work does not aspire to scientific eminence, it is hoped that, from an ethnological point of view, it will not be without interest and value'.[73]

The People of India was certainly seen in an ethnographic light. Sir John Lubbock drew upon the work in his book *The Origin of Civilization and the Primitive Condition of Man* (1870).[74] Moreover, many of the photographs reached a wider audience when they were part of a collection loaned to the ESL by Dr Forbes Watson with the encouragement of T. H. Huxley in 1869.[75] Exhibited together with maps, one of which showed the worldwide distribution of 'primeval races',[76] the photographs aided the ethnological and geographical fixing of lower races described in a series of lectures on the ethnology and anthropology of India.[77] *The People of India* collection was also shown in the British exhibit at the International Geographical Congress in Paris in August 1875, where it was judged to be 'more picturesque than scientific' but nevertheless of considerable ethnological value and consequently worthy of a second-class medal.[78]

The significance of *The People of India* lies in the ways it constructs knowledge of racial 'types' relative to their political co-operation within the administrative frameworks of colonial authority. Many photographs depict individuals posed rigidly against blank backdrops while others show groups arranged in more casual attitudes with cultural artefacts. The photographs are numbered and classified by printed labels, identifying the name, tribe, religion and place of those pictured. The terms of the racial classification vary, frequently confusing tribe, race and caste. Nevertheless, the often detailed written descriptions accompanying each photograph prescribe specific ways of reading the image as a map of physical and, by association, moral character. The text accompanying 'Jats. Hindoos. Delhi. No. 192' (illus. 59), for instance, notes that Jats have 'fine soldierly figures, and strong, but simple features'. Physiognomy and gender are thus read in utilitarian ways. Noting that the Jats remained loyal during the 'Mutiny', they are described as 'a fine, manly race of men [making] good steady soldiers'.[79] The Pachada tribe, in stark contrast, is represented in a single 'type' (illus. 60) said to exhibit an hereditary tendency to 'rapine, murder and robbery on a large scale'. Reading the photograph in close detail, the text continues, 'His countenance is extremely forbidding, and is perhaps an index to the lawless and unreclaimed nature of his tribe.' Thus the photograph and text construct the Pachada tribe as a criminal type

59 'Jats. Hindoos. Delhi. No. 192', J. Forbes Watson and J. W. Kaye, *The People of India* (1868–75).

60 'Hoosein. Pachada, Formerly Hindoo. Hissar. No. 180', in J. Forbes Watson and J. W. Kaye, *The People of India* (1868–75).

whose hereditary lawlessness may be read through physiognomy.[80] The reader is warned: 'These and people like them, are bad elements in the general population of the North-West provinces, and need constant watching.'[81]

In this way, types of people could be compared and contrasted and placed within the structure of imperial administration and rule. Such photographs, and the project of which they were part, may be read as a form of imperial surveillance; part of the imperial desire for the total visibility of peoples and places.[82] Other forms of colonial survey, such as the first census of India in 1881, similarly attempted to comprehend the 'condition' of Britain's Indian empire, by ordering the 'population' within a regularized framework.[83] Such modes of survey were central to Orientalist systems of knowledge; 'Western techniques of representation that make the Orient visible, clear, "there" in discourse about it'.[84] India in this way becomes a vast geographical museum, a space for Western knowledges to construct 'types' of people in their proper places.

Surveying African 'Types'

While photography was used in large official ethnographic projects such as *The People of India*, it was also used by individual imperial travellers, writers and administrators. The colonial administrator, geographer and naturalist Harry Hamilton Johnston (1858–1927), for instance, started using photography extensively as a tool of scientific research and a means of advocating colonial expansion in Africa while serving as Commissioner for British Central Africa in the 1890s.[85] His illustrated publications, such as *British Central Africa* (1897), used photographs to prove his theories of the low intellectual, moral and cultural status of 'the negro race', and to provide evidence for its improvement under British colonial rule.[86]

Johnston's own official authority in Africa was at its zenith from 1899 to 1901, when he was Special Commissioner to the Uganda Protectorate, during which time he amassed a collection of almost 1,000 photographs, which he used in numerous lectures and publications.[87] The Commission, in which Johnston worked closely with W. G. Doggett, taxidermist and photographer, set about examining the Uganda Protectorate with colonial eyes: studying railway routes, settling questions of native taxation, and formulating a policy on land ownership. As well as collecting zoological, geological and botanical specimens and ethnological objects for the British Museum, Johnston and Doggett also undertook an ethnographic survey of the 'marked types of African Man', from 'the low ape-like types of the Elgon and Semliki forests' to the 'Apollo-like Masai'.[88] Johnston used photographs of 'types' to place different groups both in space and on an evolutionary scale. Johnston described the 'Andorobo' in Uganda's eastern province, for example, as a 'mongrel

61 Harry H. Johnston, 'Andorobo drinking: Molo River', *c.* 1900.

nomad race', picturing them in carefully posed photographs with titles such as 'Andorobo drinking: Molo River' (illus. 61) in order to prove that they offer 'a striking illustration, handed down through the ages, of the life of primitive man not long after he had attained the status of humanity'.[89]

Johnston's photograph showing 'Doggett and Muamba' (illus. 62) is not only an image of anthropological activity. In portraying a profile view of an African man's upper body, the photograph also conforms to established rules of anthropometric photography. The measuring and photographing of 'racial' types by the Special Commission were indeed designed to comprehend different peoples within a scientific framework. In this case, however, the scientist himself, rather than an anonymous measuring screen or rod, is shown in the process of calculating anthropological statistics of physical difference. Yet by establishing his presence the anthropologist here has the same iconic function as the measuring screen. Ironically, in the process of measuring a 'native type', Doggett has been represented as also caught between the callipers of imperial science. Indeed, the photograph is itself a measure of the imperial anthropologist and his presumed superiority to those he represents. Johnston's photographs of 'native types', were intended ultimately to prove the African's

62 Harry H. Johnston, 'Doggett and Muamba', *c.* 1900.

racial inferiority, though to different degrees, to the European. His use of photography in this particular example was a singularly important means of demonstrating that Uganda was the 'White Man's Colony'.

Commercial 'Types'

Commercial photographers also saw their subjects through the lenses of anthropological inquiry and the iconography of 'race'. This was especially true of photographers like John Thomson, who was a Fellow of the ESL and was keen to make his work useful to scientific studies of race. Following his use of photographs, along with descriptions of muscular development and estimates of average heights, as part of his report on the ethnology of Cambodia to the ESL in 1867, he took particular interest in his portrayal of human subjects.[90] His massive 1873–4 work, *Illustrations of China and Its People*, followed the work of other European travellers and residents such as the Reverend Justus Doolittle whose images of Chinese people and customs had been published in 1868.[91] Thomson also adopted an already well-established system of representing different people, through the discourse of 'racial types'. As one enthusiastic reviewer of his *Illustrations of China and Its People* noted, 'the photographs are exceptionally valuable on account of the types of human form and character depicted'.[92] The reviewer singled out for particular commendation a plate depicting 'Male Heads, Chinese and Mongolian' (illus. 63). Thomson's text prescribed a particular way of reading the photographs. No. 20, according to Thomson, represented a boy from 'the upper or most highly educated class, the son of a distinguished civil officer of Canton'.[93] Although 'a fine, attractive-looking little fellow, his full hazel eyes beaming with kindliness and intelligence', the boy's face would, Thomson argued,

> gradually lose its attractions as it grows to maturity. The softness of the eye is then frequently replaced by a cold, calculating expression, the result of their peculiar training, and the countenance assumes an air of apathetic indifference which is so necessary to veil the inner feelings of a polished Chinese gentleman.[94]

No. 21, according to Thomson, represented 'the head of a full-grown Chinaman' and what the boy in no. 20 'may in time become': a man with 'natural shrewdness and capacity for business'. No. 23 is said to represent an old Chinese 'labourer', while no. 24, Thomson notes, represents

> the head of an ordinary Chinese coolie, a fine specimen of the lower orders in China . . . He is, as a rule, a kindly-disposed person, quite alive to his own interests, and endowed by nature with a profound contempt and compassion for all barbarians who dwell without the pale of Chinese civilization. This will account for the expression he is casting upon me as I am about to hand him down to posterity to be a type of his class.[95]

Thomson was indeed engaged in a project of racial classification, and his photographs of people are constructed around the common ethnological

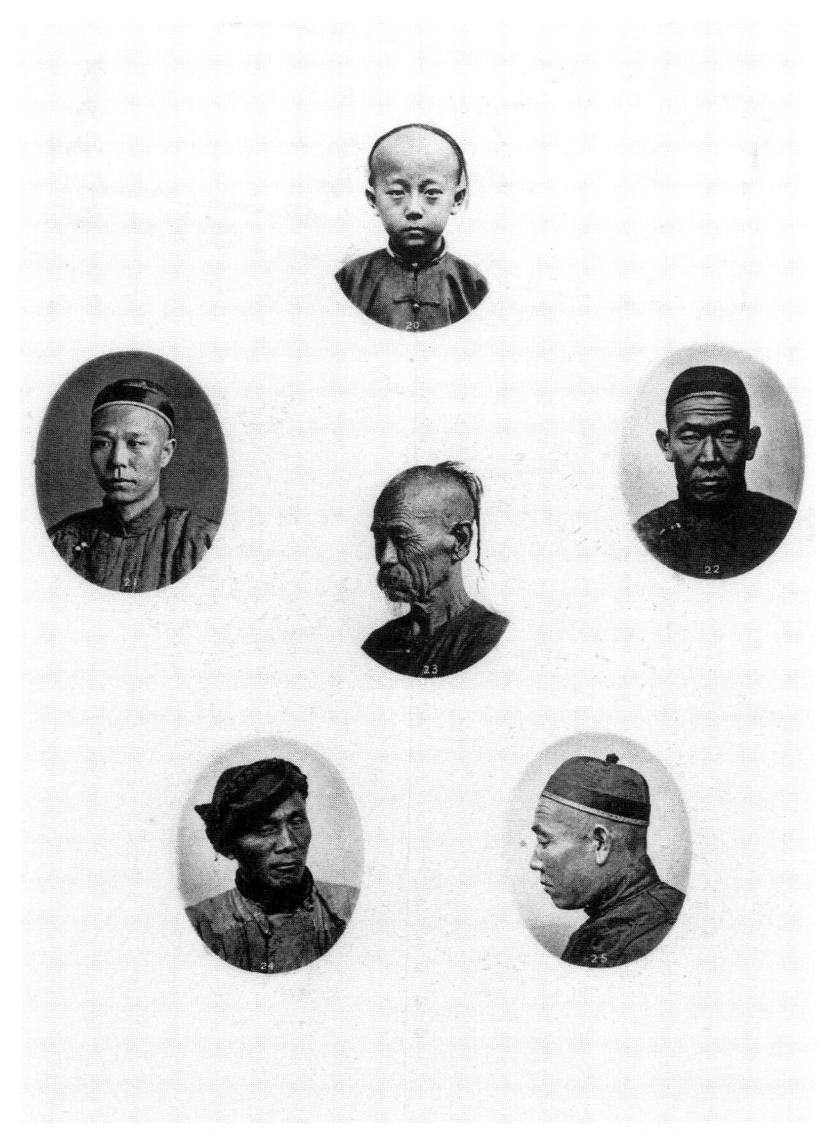

63 John Thomson, 'Male Heads, Chinese and Mongolian', *Illustrations of China and Its People* (1873–4).

concept of racial 'types'. His numbered photographic vignettes of types and their accompanying texts are reminiscent of the iconography and rhetorical style of *The People of India*. Indeed, both works were displayed as part of the British exhibit at the International Geographical Congress in Paris in 1875, where they both won second-class medals.[96] Like the contributors to *The*

People of India, Thomson classified people visually, constructing knowledge of their physical and moral character. The latter was often evaluated in terms of usefulness for British commerce and the Western traveller.

Thomson's use of full-face and profile views of isolated, individual heads, such as nos. 22 and 25, which represent 'a Mongol' whose 'features are heavier than those of the pure Chinese', is clear evidence that he was influenced by conventions of anthropological photography and phrenology.[97] He later used photographs to support the racial classification of the 'Mongol tribes' in terms of their 'high and broad shoulders, short broad noses, pointed and prominent chins, long teeth distant from each other, eyes black, elliptical, and unsteady, thick short necks . . .'[98] He also later advocated a more systematic application of photography to anthropology, noting in 1891 that in the photographing of a racial type 'it is necessary to take a full face and profile view of the head, or a series of overlapping views of a number of types of the same family'. He also advocated the inclusion of a measuring rod in the photographs of individual heads and full-length figures to 'give a basis of measurement'.[99] While he did not apply any systematic anthropometric methodology to his own photography, such as that devised by J. H. Lamprey, Thomson nevertheless pursued forms of racial classification through a less formal style.

Like many commercial photographers, Thomson represented racial 'types' not just through individual appearance or parts of the body. He also photographed people engaged in particular occupations in scenes of 'street groups', particularly in major cities. His 'Street Groups, Kiu-kiang' (illus. 64) shows four distinct groups which he identified (from left to right) as 'Ahong . . . a maker of soup' with customer, a 'public scribe' writing a letter at the dictation of a woman customer, an 'itinerant barber' and, lastly, a 'wood-turner and his customer' who is examining 'a wooden ladle'.[100] He provided intimate details about the main characters and apparatus of each group, from the life history of 'Ahong' the soup maker to the contents of the drawers of the cabinet upon which the barber's client sits. Thomson's verbal and pictorial descriptions are also moral evaluations. The 'public scribe', who works also as 'a fortune-teller and physician', is 'a crafty old rogue, and trades on human credulity with astounding success'. Thus while the photographs of individual heads tend to categorize through facial appearance, these scenes picture 'types' through a sense of their 'natural' trades and occupation.

Because of the fear and hostility expressed towards Thomson and his photographic apparatus, and because of the formalities unavoidable in his wet-plate photography, such studies were undoubtedly carefully posed with, it would appear, paid participants. Thomson undertook many photographic studies of individual street groups, from 'Chinese Medical Men' to 'Dealers in Ancient Bronzes'.[101] However, 'Street Groups, Kiu-kiang' is particularly significant for its combination of a number of groups into one scene. It is an

64 John Thomson, 'Street Groups, Kiu-Kiang', *Illustrations of China and Its People* (1873–74).

attempt to capture in a single photograph a diorama of 'types' of street groups and their trades.

Even before the advent of photography explorers had deployed such a technique in their ethnographic descriptions. For example, describing 'The Palaver' from his *Picturesque Views on the River Niger* (1840), William Allen noted: 'I have grouped together all the principal characters of whom I had individual sketches. They are, I believe, likenesses.'[102] This technique also finds echoes in later anthropological photographs which represent a range of different cultural activities in a single image, such as that by E. H. Man showing 'Andamanese Shooting, Dancing, Sleeping and Greeting' (*c.* 1880).[103] It also has strong parallels with the display of social types in Victorian narrative art. For example, paintings such as William Frith's *Derby Day* (1858) and *Railway Station* (1862) presented complex social scenes, with a range of 'types' whose clothes, faces, figures, and gestures required reading with anthropological scrutiny.[104] Indeed, like a number of contemporary artists, Frith actually used photographs to aid his delineation of the human social type in his paintings.

Thomson was not alone in his efforts to use photography as a means of representing the ethnological features of China. The traveller, geographer

65 John Thomson, 'The Cangue', *Illustrations of China and Its People* (1873–4).

and orientalist E. Delmar Morgan made a similar series of photographs in China in the mid-1870s featuring individuals and groups posed in makeshift outdoor settings.[105]

Such photographs of Chinese 'types' or 'typical scenes' reflected established Western images of China, propagated by other travellers, writers, artists and photographers.[106] Thomson's photographs of, for example, a Chinese woman's fashionably deformed foot[107] or people smoking opium[108] confirmed China's low place on the scale of civilization and provided additional evidence to support a dominant Western image of the Chinese as hopelessly addicted to a range of vices.

Thomson's photograph of a man in a 'cangue' (illus. 65), for example, appears in *Illustrations of China and Its People* accompanied by a description of Chinese punishments. In using this image Thomson was contributing to a well-established European fascination with Chinese punishments and tortures, reinforcing prevalent Western views of the Chinese as 'masters of refined cruelty'.[109] More generally this image represents a metaphor for the perceived situation of China and the 'Chinaman': locked in uncivilized customs and burdened by poverty and ignorance. Such an image cast encouraging light on British imperial incursions within China. Used as a frontispiece to Thomson's popular book *The Straits of Malacca, Indo-China and China* (1875),[110] this image becomes the central metaphor of Thomson's conclusions on China and the Chinese:

66 John Thomson, 'A Pekingese Coster-monger', *Illustrations of China and Its People* (1873–4).

> The picture is at best a sad one; and though a ray of sunshine may brighten it here and there, yet, after all, the darkness that broods over the land becomes but the more palpable under this straggling fitful light. Poverty and ignorance we have among us in England; but no poverty so wretched, no ignorance so intense, as are found among the millions of China.[111]

Thomson here not only constructs China and the Chinese as a 'picture' to be observed and assessed in its entirety. He also employs an iconography of light and dark, making the foreign settlements and treaty ports into a 'ray of sunshine' and casting himself and his photography as an agent of Western enlightenment.

Thomson's comparison between social evils in China and those in England is also worth noting, for throughout his work 'oriental types' were explicitly compared with 'types' of urban poor in Britain. Thus he described a fruit-seller in Peking whom he photographed in the pose of going about his everyday business (illus. 66) as 'a type of a Pekingese costermonger, one of the lower orders'.[112] In making 'oriental' peoples analogous with London's poor, Thomson was drawing not just on his own experience but on a well-established literary and pictorial convention of social exploration. Indeed, Thomson's language and imagery here directly echo those of Henry Mayhew's *London Labour and the London Poor* (1861), which identified costermongers as a distinct race with their own physical appearance, habits and language. Mayhew's writing and John Beard's daguerreotypes, upon which the engravings for *London Labour* were based, indicate the beginnings of social explorations within Britain which employed the camera. Thomson would in due course contribute to this practice with his famous photographs of 'street types' in *Street Life in London* (1878–9), which I will consider in more detail below.

Thomson was by no means alone in fashioning himself as an explorer setting out to shed light on the dark recesses of social life in Britain. For instance, the photographer P. H. Emerson (1856–1936) embarked on a series of explorations of the Norfolk and Suffolk Broads in the mid-1880s during which he made picturesque studies of landscapes and the 'natives'.[113] In such pursuits, commercial photographers tapped into a market for portraits of picturesque domestic subjects, from fishermen to street urchins. Much of this portrait work was made and read through contemporary codes of portraiture and delineations of character in both science and art. The output of commercial studios was also drawn upon by individuals and institutions in the course of scientific researches on human character and classification in Britain.

Race at Home

Issues of 'race' within Victorian Britain attracted attention to the unfamiliar at home as much as to the exotic abroad. In one sense, the Victorian public did not have to travel far to encounter the racially exotic, since the popular display of non-European people live at exhibitions in Britain throughout the Victorian era provided many opportunities to picture 'types' of race. Such displays also highlighted ideas about the racial origins of the British themselves. Indeed, theories and images of 'race' were concerned as much with the British and Europeans as they were with non-Europeans. In the 1870s and 1880s, for example, the *Journal of the Royal Anthropological Institute* exhibited a 'remarkably Europocentric and even Anglocentric focus'.[114] Contemporary interest in the origins and heritage of the British national character, as well as concerns about urban degeneration, contributed to the processes by which cameras were being focused on different social groups and 'races' within Britain.

One of the most systematic attempts to use photography in the classification of racial 'types' within the United Kingdom was that of the Anthropometric and Racial Committee appointed by the BAAS in 1875 'for the Collection of Observations on the Systematic Examination of Heights, Weights &c., of Human Beings in the British Empire, and the Publication of Photographs of the Typical Races of Empire'.[115] Despite its initial imperial scope, the committee limited itself to the United Kingdom. It regarded its labours as essential for the 'elucidation of a subject which is of great national importance as well as of scientific interest'.[116] There was indeed considerable interest in the question of 'national character', especially in confirming the natural supremacy of the 'Anglo-Saxon' or 'Teuton' race. This race, it was argued, had conquered Britain, displacing the Celts and other ancient Britons into the peripheries of the British Isles, and remained the dominant race. The construction of the superior 'Anglo-Saxon' race was significant both in legitimating the dominance of the English within Britain and in giving racial justification to imperial expansion.[117]

By 1880, some 24,000 physical observations had been amassed by the BAAS, yet only some 400 photographs had been collected. This was clearly inadequate, as J. Park Harrison noted: 'A larger number of portraits, taken on a uniform system, in profile and full face, would be required, together with head-measurements, to enable the committee to define racial characteristics.'[118]

In 1882 a newly established 'photographic committee'[119] declared its faith in the scientific value of photography in representing 'a clear definition of racial features' and in the relevance of such work to a number of 'social questions' concerning the 'tendencies and proclivities' of classes as well as races.[120] It was suggested, for instance, that the work might aid the 'exact description of criminals and deserters; resulting, it cannot be doubted, in more frequent arrests'.[121] Thus the delineation of national racial 'types' was closely linked to the categorizing of particular social groups in Britain.

The targets of the committee's photographic work were intended to be more or less pure racial 'types'. Despite the wide variations in the population of Britain it was possible, the committee argued, to distinguish 'original or main racial characteristics'. Thus in the early years the photographic collection reflected a distinctly local focus, with photographs being gathered from the remoter areas of Britain. However it was found that by themselves photographs were of little use in actually defining racial characteristics. Consequently, drawing on J. B. Davis and J. Thurnam's *Crania Britannica* (1865), the committee correlated measurements taken from ancient skulls with those from living individuals, identifing three main types: A (Dolichocephalic Dark); B (Brachicephalic Fair) and C (Sub-Dolichocephalic Fair). The photographs were rearranged according to these three racial types identified by craniometry and a series of 'typical features'.[122] 'Griffith Llewellyn at 50 . . . A-Type' (illus. 67) is from one of three albums (labelled Type A, B and C respectively) compiled by J. P. Harrison for the committee.[123] The front and profile views of an individual head framed by details of the subject's age, physical appearance, occupation and locality placed the subject firmly within the category of 'type A'. Typical A features included a vertical, square forehead; straight, long nose; thick, unformed lips; narrow jaw; dark eyes; very dark, crisp, curling hair; and an average height of five feet three inches.

By contrast, type C, regarded as 'a correct definition of true Saxon features', was defined largely by the absence of the supposed defects ascribed to type A or B. Thus while type C was described as physically 'stout, well-covered', type A was 'slight' and type B was 'bony, muscular'. Similarly, the lips of the mouth in type A were described as 'thick, unformed', those of type B as 'thin, straight, long', whereas those of type C were 'well formed'.[124] There

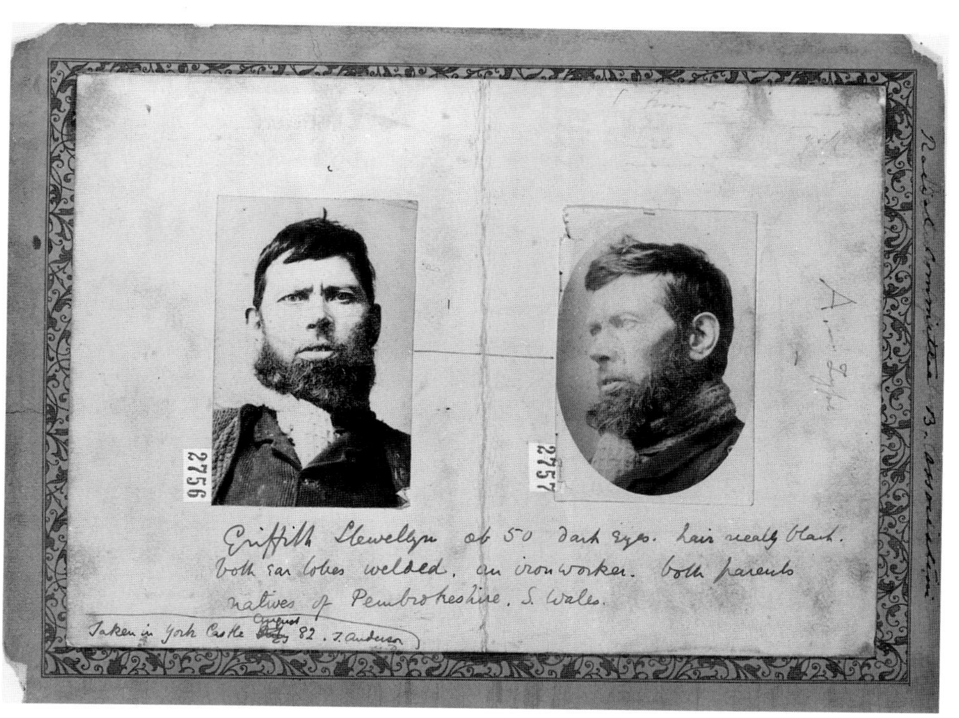

67 J. H. Anderson, 'Griffith Llewellyn at 50 . . . A-Type', 1882.

was also a geography to this racial classification. Thus, when reporting on its researches into type C 'true Saxon features' in 1882, the committee claimed, 'Photographs conforming in all respects to the above characteristics have been obtained from Sussex and several other English counties.'

Although the photograph 'Griffith Llewellyn at 50 . . . A-Type' was made specifically as an anthropometric 'type', most were selected from the *carte-de-visite* selections of commercial operators and were therefore shaped by contemporary fashions in portraiture and commercial photography. This is particularly noticeable in the photographs which place their subjects against scenic backdrops. A series of portraits of Yorkshire fishermen from album B, for example, reveals little purely anthropometric data.[125] Such figures were popular characters for Victorian studio photographers, who responded to a wider middle-class demand for images of labour and picturesque types.[126] To those who made and purchased these studio portraits, the characters are identified as 'types' not only by their physiognomy but also by their clothes, occupation and painted backdrops.

Although such photographs themselves provide little anthropometric information, they allowed the visualization of different racial types within Britain and tapped into popular visual iconographies of 'character'. John

Beddoe used a number of the BAAS photographs as the basis for illustrations of 'types' in his *Races of Britain* (1885),[127] where he reaffirmed the racial dominance of Anglo-Saxons, occupying England and the east of Scotland, with the Celtic races having been displaced to Ireland, west Scotland, Cornwall and Wales. He argued that his measurements of physiognomy and colouring correlated directly with mental and moral characteristics. Thus, for example, 'educated Englishmen' were said to have the biggest heads, while the Irish had the smallest.[128]

Such arguments concurred with the BAAS's final report of 1883 based on some 53,000 observations on the physical characteristics of the British population.[129] It was deduced, for example, that for adult males the Irish, Scots and Welsh occupied the extremes of weight and strength and height, while the 'normal' middle ground was held by the English. As in the photographs, physical features were taken as accurate indicators of 'mental' qualities. The committee's ranking of British adult male stature and weight, for example, ranged from the 'Metropolitan police', near the top, to 'idiots and imbeciles' at the bottom.[130]

While the bulk of the committee's work focused on types within Britain, it inevitably touched on wider discourses of 'race' and 'types'. For example the committee ranked the average stature of the different races and nationalities of the world, including 'Negroes of the Congo', 'Irish – all classes', 'Hindoos' and 'Andamanese',[131] noting

> It is interesting to find that, with the exception of a few imperfectly-observed South Sea Islanders . . . the English Professional Classes lead the long list, and that the Anglo-Saxon race takes the chief place in it among the civilized communities.[132]

Like the photographs collected by the committee, the statistics were said to 'speak for themselves', yet both represent prominent concerns with a decline in British men's physical and mental character.[133] This concern with racial fitness was also a major consideration behind other attempts to photograph social and racial characteristics in Britain, including the use of composite photographs.

Composite photographs

One of the prime movers in the BAAS committee was Francis Galton (1822–1911). Best known as the inventor of eugenics, Galton's researches into heredity were part of an enormous range of scientific activities and interests, including the use of photography to measure and represent human 'types'.

Galton began his experiments into the scientific applications of photography in 1865, with help from his cousin Robert Galton, promoting a portable, stereoscopic 'photographic map' for explorers and military commanders.[134] While involved in the BAAS Racial Committee in 1877 Galton suggested the

Their attempt to catalogue 'true types' situates *Street Life in London* in the tradition of ethnological depiction which structured much of the representation of the contemporary social scene within Britain as well as on its imperial peripheries. By claiming to document reality, the photographs and text naturalize the categorization of individuals within various urban 'types' occupying particular urban spaces.

Thomson's focus on 'street life' was influenced by the technical limitations of photography, and he made few pictures of the interiors of homes or workplaces. It was not until the 1880s that improved film technologies and magnesium flash-powder enabled the dark and windowless interiors of such dwellings to be photographed easily.[165] However, 'street life' was also a social category, a way of seeing and understanding cultural difference. Individuals were defined as 'types', placed in particular 'classes' and 'races' through their appearance, habits and occupation (or lack of occupation) as well as their location (or lack of location) within the city's spaces.

Street Life in London claimed, with some justification, originality in its extensive application of photography to outdoor social exploration. Although Mayhew had used Beard's daguerreotypes for his engravings, mid-Victorian photographers had showed little enthusiasm for embarking on explorations into the urban unknown compared with writers of fiction and journalism. However, studio portraits of urban poverty, emulating other images in magazines and papers, were common.[166] The commercial photographer Oscar G. Rejlander (1813–75), for example, made a number of photographs of 'ragged children' in the late 1850s and 1860s. Unlike Thomson, Rejlander was exclusively concerned with photography as a practice of fine art.[167] He posed hired child models as 'street urchins' against murky studio backdrops to present a sentimentalized and innocent view of urban poverty. The narratives these popular images constructed were avowedly fictitious; the themed portraits with titles such as 'Charity' focused more on acts of giving than those of receiving.[168]

Although Thomson's photographs avoid the overtly sentimentalized or humorous themes acted out in studio portraits by Rejlander and others, *Street Life in London* was built on the same conventional iconography of urban poverty. Thomson's photograph of 'The "Crawlers" ', which shows a woman too exhausted even to beg huddled together in a dark doorway with a child, closely resembles Rejlander's famous studio portrait from the early 1860s titled 'Night in Town' which shows a young boy, head bowed, dressed in torn and crumpled clothes, crouched in a dark doorway. 'Night in Town' was exhibited several times in London in the 1860s and was also used as a presentation print by the South London Photographic Society.[169] *Street Life in London* might have moved from the space of the studio to the space of the street, but Thomson's images were as formally composed, many having been arranged by appointment, as any studio portrait. *Street Life in London* thus often presented a highly conventional vision of urban poverty; one designed as much for aesthetic enjoyment as for social enlightenment.

Yet by claiming to document reality, Thomson's photographs and text naturalise the categorization of individuals within various urban 'types' occupying particular urban spaces. *Street Life in London* presented its middle-class purchasers with an intelligible social topography, from the flower girls of Covent Garden to the 'Crawlers' of St. Giles.

While *Street Life in London* drew on and added to existing iconographies of urban characters, the work also made explicit analogies between on the one hand non-European uncivilized 'types' and landscapes and on the other those unfortunate urban 'types' lost in the midst of civilization, hidden in the dark spaces of the city. Beginning the story accompanying 'London Nomades' (illus. 69), Thomson commented:

> In his savage state, whether inhabiting the marshes of Equatorial Africa, or the mountain ranges of Formosa, man is fain to wander, seeking his sustenance in the fruits of the earth or products of the chase. On the other hand, in the most civilized communities the wanderers become distributors of food and of industrial products to those who spend their days in the ceaseless toil of city life. Hence it is that in London there are a number of what may be termed, owing to their wandering, unsettled habits, nomadic tribes.

The people arranged and photographed around the caravan of one William Hampton, on some vacant land at Battersea, were described as 'Nomades', as part of a 'tribe' of London poor of which Hampton, according to Thomson, represented 'a fair type'. As such, Thomson explicitly compared Hampton and his associates to 'the Nomades who wander over the Mongolian Steppes, drifting about with their flocks and herds, seeking the purest springs and greenest pastures'.

Thomson is here emulating Mayhew, who presented himself as a 'traveller in the undiscovered country of the poor'.[170] Just as Mayhew had divided the world broadly into 'the wanderers and the settlers – the vagabond and the citizen – the nomadic and the civilized tribes', each with distinctive moral attributes,[171] Thomson described the 'London Nomades' as 'improvident' and 'unable to follow any intelligent plan of life'.[172] His classification of 'types' was also based on physical appearance and in comparing 'London Nomades' explicitly to Mongolian tribes, he was referring not merely to cultural habits. Indeed, his descriptions of the nomadic 'Mongolians' ('high cheekbones [are] thoroughly characteristic of the Mongol race'[173]) correlate closely with Mayhew's assertion that nomadic tribes have 'high cheek-bones and protruding jaws'.[174]

In likening groups of London 'street types' to Oriental tribes, Thomson was also extending his earlier categorizations of Chinese street types, such as the

69 John Thomson, 'London Nomades', *Street Life in London* (1878).

'type of a Pekingese Costermonger' (see illus. 66), which drew in turn on Mayhew's categorization of 'Costermongers' as a distinct race whose notions of morality 'agree strangely ... with those of many savage tribes ... they are part of the Nomades of England, neither knowing nor caring for the enjoyments of the home'.[175]

In collapsing the outside 'unknown' world on to the space of the imperial metropolis, Thomson drew on a way of seeing, fostered by Mayhew and others, which viewed London as a 'distinct world ... composed of different races like a world'.[176] In this way social explorers like Mayhew and John Binney could employ the language of imperialism in noting that localities 'in

70 Arthur Concanen, 'A Visit to "Tiger Bay"', in James Greenwood, *The Wilds of London* (1874).

the heart of the City, are as much *terra incognita*, to the great body of Londoners themselves, as Lake Tchad in the Centre of Africa'.[177] The 'world of London', they argued, presented almost every geographical species of the human family: 'If Arabia has its nomadic tribes, the British Metropolis has its vagrant hordes as well. If the Carib Islanders have their savages, the English Capital has types almost as brutal and uncivilized as they.'[178]

Journalistic guides to 'Darkest London' such as James Greenwood's *The Wilds of London* (1874), to which Thomson and Smith acknowledged their debt, also drew on the same associations of 'savagery' in the metropolis and on the imperial frontier.[179] Greenwood's writings, like Thomson's photography, easily straddle such worlds. His studies of 'Darkest London' and the 'Great Residuum' at home shared the imperial rhetoric and ethnological iconography of his adventure books such as *Prince Dick of Dahomey* (1890) and highly sensationalist works such as *The Wild Man at Home*.[180]

One of Greenwood's early explorations in *The Wilds of London* leads him to 'Tiger Bay', a notorious part of the East End where sailors were lured and robbed by unscrupulous women ('tigresses'). Arthur Concanen's drawing of 'A Visit to "Tiger Bay"' depicts a scene in a public house with the 'tigresses' dancing and drinking with, and stealing from, drunken sailors (illus. 70). It presents Greenwood's observation of this 'spectacle' as if '"behind the scenes" at a theatre during the pantomime season'.[181] The detailed drawing shows clearly the physiognomy of both sailors and women. Greenwood's description of the 'tigresses' reinforces the ethnological imagery: 'The same short, bull-like throats, the same high cheek-bones and deep-set eyes, the same low retreating foreheads and straight wide mouths, and capacious nostrils, the same tremendous muscular development stamps one and all.'[182]

This description echoes Francis Galton's account of the similar appearance of 'Hottentots' in Africa and the 'felon face' of 'bad character' in England. Indeed, both are closely associated with the language of phrenology and codes of caricature which mark out Concanen's figures as 'bad characters'. The prominence of the black man in the image is also notable, since blackness was frequently used as a sign of physical and mental degeneracy and sexual excess.[183] Scientific inquiries such as those carried out by the BAAS in the 1880s attempted to measure the degeneration of social groups by calculating their 'degree of nigresence'. Thus criminals were shown to have an 'excess' of 10 per cent of dark eyes combined with dark complexion over the general population.[184] The intimacy of 'degenerate' black with 'degenerating' English women, as shown in Concanen's picture, reinforced assumptions as to the inferiority of both in relation to English men, further conflating discourses of degeneracy in race and gender.[185]

With his language of racial 'types' and rhetoric of 'exposing . . . those hole-and-corner evils which afflict society',[186] Greenwood's explorations were metaphorically 'photographic'. Indeed, photography served as a powerful metaphor in practices such as journalism for complete, accurate exposures of the unknown through first-hand observation and interviewing. Mayhew's *London Labour and the London Poor* was thus put forward as 'a photograph of life as actually spent by the lower classes of the Metropolis'.[187] Other social explorers such as E. C. Grenville Murray and writers such as Charles Dickens similarly employed photography as a metaphor.[188]

Although Thomson used photography practically, and not simply metaphorically, in many ways he followed routes already established by earlier social explorers. Although an essential characteristic of 'London Nomades' was that they were 'unsettled', constantly on the move between places, many of Thomson's photographs locate 'street life' in south and east London, with some specifically in the East End.[189] Mid-Victorian social explorers, notably Jerrold and Doré in *London: A Pilgrimage,* had increasingly contrasted London's East End, as a place of poverty and vice, with its glamorous West End.[190] Furthermore, this contrast had wider connotations as writers like Mayhew and Binney presented London as a model of the globe and an 'aggregate of *various nations*'.[191] In this reading, Hampstead and Sydenham became north and south poles, 'with the whole line of Oxford Street, Holborn, and Cheapside scorching under the everlasting summer of what would then be the metropolitan torrid zone; and while it was day at Kensington, night reigning at Mile End'.[192] Through descriptions of degenerate spaces such as opium dens, social explorers such as Jerrold and Doré orientalized the East End of London.[193] Similarly, Greenwood's description of East End women as 'tigresses' represented these 'types' of women and the urban spaces they occupied as sites of unrestrained orientalized sexuality.[194] In his explorations of China and London, Thomson was reinforcing such imaginative processes whereby the Far East was mapped on to the East End of London; both became areas of darkness, danger and the 'unknown', to be explored and subjected to scrutiny.

Multiple 'Types'

In photographing 'the natives' in the Empire and within Britain, photographers were simultaneously representing a portrait of their own culture. Such features were structured through a predominant concern for delineating the physical appearance and moral proclivities of different 'races', and yet it is hard, if not impossible, to identify the typical anthropological photograph, since almost any photograph might be classified within an anthropological frame.

In the representation of 'types' considered here, it was commonly assumed that external, physical appearance was a key to internal, moral constitution. Photographs reinforced wider racial categories through their presentation as objective and scientific records. In this way the technique of photography and the concept of the 'type' were together given a central place within projects of human classification. The camera, then, was just one instrument within an armament of strategies of surveying and measurement designed to produce scientific knowledge of different 'physical characters', playing a central part of the processes of recording and classifying which made foreign peoples visible and understandable.[195] The practice of photography within a discourse of anthropology was also consistent within its wider cultural role to maintain the reality of exotic and foreign worlds while keeping them separate and distanced from the familiar world at home.[196]

Yet, as I have suggested, the 'type' was not limited to categorizing non-Europeans but structured complex hierarchies within British society itself. Moreover, the representation of 'types' within Britain was bound up with the representation of those outside Britain. Thus the idea of the Negro 'type' constructed by Harry Johnston was determined by his idea of their place within a world dominated by a superior Anglo-Saxon and Caucasian race. As a result, in his *The Negro in the New World* (1910) he likened the 'Negro race' to other 'primitive' races, comparing their photographs to one of the 'Caucasian type' represented by an Englishman.

The broad currency of the 'type' across discourses of both art and science meant that photographs could signify as 'types' in a whole range of ways. Attempts to use photography within anthropometry were based on an idea of the racial type as an accurate, comparative measurement of human difference.

71 'Some of our Pacific Island cousins and their home. Tutuila, Samoan Is.', stereoscope slide, *c.* 1890.

Galton's researches with composite photography developed an idea of the type in almost statistical terms, calculating averages and variation. In reading the *cartes-de-visite* collected by the Racial Committee of the BAAS, on the other hand, clear physiognomic differences are far less legible than the signs of 'character' indicated by props, dress, gender and occupation.

Thus, despite claims to scientific certainty, photographs of 'types' were not always seen to provide conclusive proof of the correlation between appearance and moral character. As anthropologists became more interested in the lived culture of different races, they began to question the usefulness of anthropometric photographs of racial 'types'. In 1893 Everard im Thurn criticized the 'purely physiological photographs of the anthropometrists' as 'merely pictures of lifeless bodies'. He also compared equally lifeless photographs of 'uncharacteristically miserable natives' with photographs of 'badly stuffed and distorted animals'.[197] Only when anthropologists and others began in this way to question the kinds of truths such observations were intended to generate did photography begin to be used in ways which dislodged the centrality of the constructed 'type'.

Commercial photographers had always been interested in more than 'lifeless bodies' and while the 'type' was to decline in importance within anthropology, a genre of the ethnographic picturesque was to remain firmly entrenched in commercial photography. This was especially true in forms of popular photographic publishing of the late nineteenth century such as lantern-slides, illustrated magazines and stereoscopic views, which, directed particularly at children, played an increasingly prominent role in teaching, especially in subjects like geography.[198] The Keystone View Company, for example, published thousands of stereoscopic views of 'types' of race for entertainment and education from the 1880s onwards. Looking at such photographs through a 'stereo viewer' produced the illusion of a three-dimensional scene, plunging the spectator into a virtual world of difference. In addition, the card's reverse carried instructive information on the geography and racial groups presented. A view of 'Some of our Pacific Island cousins and their home' (illus. 71) was thus one of a mass of such views of the Pacific produced at the time which represented the geography of Samoa as a Pacific paradise and categorized Samoans as both members of a family of 'races of mankind' and as a kind of 'noble savage'.[199] As I go on to discuss in the following chapter, images of race and place thus assumed an even greater significance as photographic technologies were incorporated into various kinds of instruction and education.

6 Visual Instruction

Pictorial Lessons

Queen Victoria's Diamond Jubilee of 1897 generated a plethora of popular photographic publications celebrating the Queen and her Empire. *The Queen, Her Empire and the English-Speaking World* (1897) and its earlier incarnation, *The English-Speaking World* (1896), for instance, each used nearly 200 half-tone photographic reproductions to convey the 'good, pure home-life at the head of the nation' as well as the scenery, cities and industry of an 'Empire upon which the sun does indeed never set'.[1] A number of publishers employed new half-tone reproduction processes to mass-produce a range of similar photographic publications, often issued initially in weekly parts, celebrating the triumphs of British imperial expansion. *Broader Britain* (1895), for example, first published as a series of twelve weekly portfolios each containing sixteen photographic views, was intended to show how Britannia's children 'have gone further afield – and yet kept home closer in their hearts – than the colonists of any other race or time'. The introduction continued: 'And it has not been without stout use of her strong white arm that she has assured to her teeming offspring peace and plenty at the ends of the earth.'[2] The 'strong white arm' of publishers and editors was certainly evident in the photographs of non-European peoples, from 'Canadian Indians' to 'Australian Aboriginals', who were consistently depicted through crude stereotypes.[3] Through their representation of subject peoples, colonial government, industrial progress and military strength, these photographic works contributed to a broader promotion of a unified Empire as the nineteenth century neared its close.

One of the most elaborate photographic surveys of Empire to be published at this time was *The Queen's Empire* (1897).[4] Issued initially in twenty-four fortnightly parts, it surveyed the Empire in nearly 300 photographs and descriptive captions. Each part covered an aspect of Empire ranging from 'The Capitals and Chief Cities' and 'Customs and Ceremonies' to ' The Education of the Empire' and 'Engineering Triumphs'. While each issue was intended to stand on its own as a slice of life throughout the Empire, the work as a whole

72 'The lowest step on the ladder of knowledge', *The Queen's Empire,* edited and with an introduction by H. O. Arnold-Forster (2nd edition, 1902).

provided an ambitious panorama which, like the coloured map of the world included in the final issue, projected Empire as a naturally united entity, centred on Britain. For in Queen Victoria's Empire, as Hugh Oakeley Arnold-Forster, MP, argued in his introduction,

> with all the variety and all the novelty, there is yet, happily, one bond of union, one mark of uniformity. In every part of Empire we shall find some trace of the work which Britain is doing throughout the world – the work of civilizing, of governing, of protecting life and property, and of extending the benefits of trade and commerce.[5]

Through careful and selective use of photographs drawn from a variety of sources, including private individuals, imperial institutions, railway companies and government departments, *The Queen's Empire* projected just such an authoritative vision of Empire.

Given its uncompromising, instructional tone, it is unsurprising that *The Queen's Empire* was keen to stress education as a key aspect of imperial unity. In a characteristic photograph from the issue on 'The Education of the Empire', bright colonial sunshine illuminates the darkness of a Queensland school (illus. 72) where, the accompanying text informs us, 'the British schoolmaster has come to the rescue with his A B C'. Here Australian Aborigines

73 'Pictures of Our Empire; no. 2: Australia' (1909).

were presented as a race on the 'lowest step on the ladder of knowledge' who might be elevated and enlightened by the lessons, language and culture of British rule. Empire was often pictured thus, as one great school in civilization, and education was a powerful means of shaping the identities and loyalties of colonial subjects at home as well as abroad. Many of these photographic publications were aimed at a wide, young audience – particularly boys – in Britain. *The Queen's Empire* was certainly intended by its editor and publisher to be educational in the broadest sense. While its various subsequent editions ensured it was popularly dispersed, a number of its images were also issued from 1907 in a series of eight large posters of 'Pictures of Our Empire', designed for the walls of schools, clubs and imperial institutions.[6] Each poster included six half-tone photographs showing characteristic scenes - in the case of Australia (illus. 73), from a sheep run in New South Wales to the Australian bush. Taken together the posters, like *The Queen's Empire* itself, amounted to a visual panorama of the natural resources, agriculture, trades and communications of Empire. As such pictures of colonial geography suggest, modes of visual instruction were intimately associated with the language and practice of teaching in schools, particularly in the subject of geography.

It was thus not only commercial publishers who regarded visual instruction, especially through photographic techniques, as an effective means of imparting appropriate knowledge of Empire to a mass audience. Educationalists, government officials and representatives of imperial propaganda societies in late-nineteenth-century Britain increasingly argued that educating the future imperial citizens was a project of crucial importance which involved a good deal more than simply 'A B C'. While school 'readers' had long been loaded with 'picture-stories' of colonial enterprise and adventure, there were increasing calls for a greater use of 'visual instruction' – through maps, models, diagrams, photographs and objects – delivered via exhibitions, 'object' lessons, textbooks and lantern-slide lectures. In what follows, I examine one of the most ambitious and intriguing official programmes of imperial visual instruction ever to have been attempted.

The Colonial Office Visual Instruction Committee

Between its foundation in 1902 and its demise after the First World War, the Colonial Office Visual Instruction Committee (COVIC) developed an Empire-wide scheme of lantern-slide lectures and illustrated textbooks to instruct, first, the children of Britain about their Empire and, second, the children of the Empire about the 'Mother Country'. The project was certainly an elaborate exercise in imperial propaganda,[7] yet to see it merely as an isolated propaganda scheme is to miss the important ways in which it articulated widespread ideas about imperial unity, citizenship and visual instruction. In

advocating their wider use in geographical teaching.[37] Lantern-slides had been used since at least the mid-1860s, along with diagrams, maps and live experiments, in the popular dissemination of science.[38] In schools the requirements of teachers for lantern-slides in geography became so pressing that they precipitated the establishment of the Geographical Association in 1893.[39] This influential association for schoolteachers began life as a means of co-ordinating the preparation and distribution of sets of geographical lantern-slides for schools subscribing to its 'lantern fund'. Within only a couple of years, at the Sixth International Geographical Congress in London in 1895, the Geographical Association could exhibit over twenty specimen sets of slides.[40] In picturing geography, many such lantern-slides were framed by imperial concerns. The Geographical Association's first President, Douglas Freshfield, thus urged the use of lantern-slides, along with maps and geographical pictures, as a means of familiarizing Britain's young with distant colonial prospects.[41] They became such a regular feature of geographical lectures that the RGS itself, despite objections from those for whom the magic lantern was synonymous with 'a Sunday School treat', purchased its own projector in 1890.[42]

The COVIC's promotion of visual instruction through lantern-slide lectures was thus part of wider currents of thought and practice within geography education. By the time he became involved with the COVIC, Mackinder had not only had close involvement with the Geographical Association but had developed considerable expertise in giving lantern-slide lectures on matters imperial and geographical. Such expertise was to prove essential in the careful preparation of pictures and text for the COVIC's scheme.

Picturing India

The major target of Hugh Fisher's first voyage in his capacity as official artist-photographer for the COVIC was India. From the hundreds of photographs Fisher sent back to London, Mackinder selected images for a set of 480 coloured lantern-slides. The lantern-slides and the eight lectures Mackinder wrote around them, completed slowly due to his other commitments and the official involvement of the India Office, were published in 1910 along with a cheaper illustrated textbook edition.[43] India occupied a central place in the British imperial imagination and its representation in these lantern-slide lectures drew on a number of historical themes, from a fascination with exotic 'types' to the display of Western technology.

In what follows I focus on the representation of India in the lantern-slide lectures, drawing upon Fisher's photographs and his diary-letters to Mackinder. Exploring such themes highlights the historical traditions upon which the COVIC depended and also exposes some of the contradictions within the COVIC project as a whole. As I have noted, Fisher's official instructions

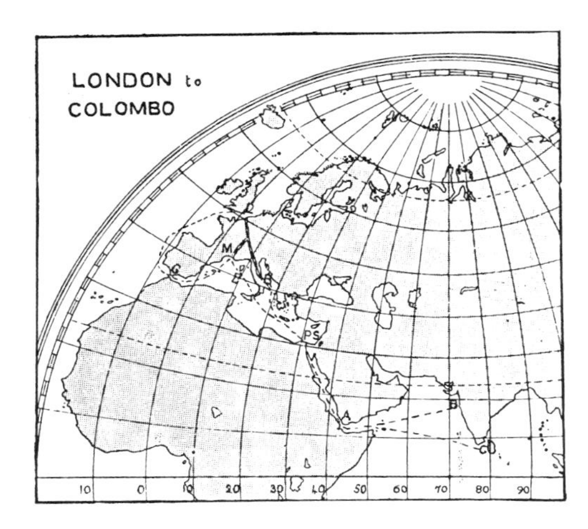

75 'London to Colombo', in H. J. Mackinder, *India: Eight Lectures* (1910).

from the COVIC requested that he record both the 'native' characteristics and the 'super-added' benefits of British rule. Out in the imperial field, however, such distinctions were not so easy for Fisher to maintain. His diary-letters to Mackinder, on which Fisher based his book *Through India and Burmah with Pen and Brush* (1910), often paint a very different series of 'travel pictures'.[44] His long descriptions and poetic ramblings were attempts, often frustrated because of his lack of time, to make a 'word picture'. These literary images were based upon a model of impressionist painting rather than photography – a medium to which he was a newcomer.[45] Indeed, both Fisher's book and his diary-letters rarely mention his photographic activities, a surprising omission given that photography was the prime focus of his travels. Despite Mackinder's aim to have one photographer documenting all the significant aspects of the Empire from a single perspective, Fisher's photographic vision was far from unitary. His photographs and writings often undermined the certainties his mission was intended to capture and display.

Mackinder's lantern-slide lectures on India transported their audiences along what Mackinder called the 'naval high street of the British Empire', from London to Colombo and then to India for an extensive tour of Britain's 'Empire within an Empire'.[46] Burma, administratively part of the Indian Empire, was also included in this tour. Maps were carefully placed to show the location and route of the visual narrative (illus. 75). Photographs projected an imaginative geography that the audiences were invited to explore through the eyes of white English men. Thus Mackinder dramatically set the scene of arrival in India:

> We land. Dark gesticulating figures surround us, scantily clad in white cotton . . . The express train to Madras is waiting, but we have a short time for that first stroll which leaves so deep an impression on the traveller setting foot in a new land. [47]

Mackinder's lectures invoked the idea of India as one vast exhibition, or picturesque spectacle, awaiting the wandering European traveller. In this they were part of a European tradition in which the Orient was experienced and represented as a picture or exhibition.[48] British exhibitions of India had, from the mid-nineteenth century, been powerful public arenas for justifying Empire and promoting imperial consolidation.[49] Images of gateways and arches, as openings into an exotic world, recur throughout Mackinder's lantern-slide lectures. One of the early images used in Mackinder's first lecture on India was a view of the Arch of Welcome in Trichinopoly, in the far south of India. The arch, upon which was inscribed 'Glorious Welcome to Our Future Emperor', had been erected in honour of the tour of India in 1905–6 by the Prince and Princess of Wales. Sidney Low covered the royal tour as a London newspaper correspondent and later tried to counter what he thought of as the British public's indifference to India in his book *A Vision of India* (1906), a 'survey' of 'our vast and varied dominion' based on his impressions.[50] Royal Empire tours were themselves exercises intended to display imperial consolidation, showing, as the editor of the *Empire Review* put it in 1901, 'beyond doubt that the British Empire is united, and that all the parts fit together into one frame'.[51] In his instructions to Fisher, Mackinder had explicitly likened the overall effect he wanted to that which might be gained by an intelligent traveller on an official tour.[52] Fisher himself was treated as an official visitor, with local officials acting as guides and procuring for him servants and transport. The image of the imperial archway, reproduced on the cover of Mackinder's textbook, also represented a central theme in his lectures: that all the variety and visual splendour of India was framed and united within the overall embrace of British rule and the authority of the British monarch.

Photographs of imperial cities and their architectural achievements were common in published surveys and celebrations of the Empire.[53] Images of monuments and memorials built under British authority in India were particularly prolific. As Fisher told Mackinder in a letter, 'The Kashmir gate must be nearly as well known in London from photographs as Buckingham Palace.'[54] In each of the main cities he visited, following Mackinder's guidelines, Fisher documented architectural icons to British rule. Thus the 'superadded' aspects of Madras were shown through scenes of Fort St George, the High Court, St Mary's Church, the Law College, the YMCA building, the Bank and the People's Park.[55]

Although Fisher followed his instructions, it seems he often found such photographic duties rather tedious. He discovered that by obtaining a high vantage point he had a particularly efficient way of encapsulating these public signs. From the top of the Ochterlony Monument, for example, like other British photographers before him,[56] he surveyed with his camera Calcutta's monuments to Empire, from Government House to the distant spire of the

77 'Baily Guard Gate', S. H. Benson, *c.* 1900.

78 'Burning Ghat, Calcutta; Bible III', coloured lantern slide from *Bible Readings*, *c.* 1900.

76 'The Burning Ghat, Benares'; 'The Baillie Gate, Lucknow', in H. J. Mackinder, *India: Eight Lectures* (1910).

English Cathedral. In his lectures, Mackinder explicitly contrasted these lofty views with an image of Calcutta's Tiretta Bazaar Street, epitomizing for him the 'Native City with its narrow ways and crowded life'.[57] Mackinder similarly used Fisher's bird's-eye views of Bombay and Delhi.[58]

Perhaps the single most important influence on the making of British images of India in the nineteenth century was the so-called Mutiny of 1857.[59] This remained a persistent theme within Fisher's photographs and Mackinder's lectures. One of Fisher's photographs depicts the Baily Gate of the Residency at Lucknow, a celebrated scene of British military defence (illus. 76). In front of the battle-marked gate Fisher pictured William Ireland, a veteran of 1857 and defender of the gate. In touring and photographing monuments such as the Massacre Ghat at Lucknow or the Memorial Well at Cawnpore,[60] Fisher was remapping and embellishing a symbolic geography of the Mutiny. Similar photographs of the Baily Gate were made by commercial lantern-slide firms, such as the London-based S. H. Benson (illus. 77), and photographic companies such as Bourne and Shepherd, and were commonly featured in publications such as *The Queen's Empire*.[61] Mackinder referred to the Baily Gate as the 'scene of the most furious attack on the British position' and used Fisher's image to testify to the courage of British soldiers in 1857 and the wide and continuing loyalty of India within the Empire.[62] His text urged the reader to be both moved by and proud of the heroic and loyal deeds which had saved the British in India and had precipitated the replacement of the East India Company with the direct government of Queen Victoria.[63]

In the textbook of Mackinder's lectures on India, the Baily Gate was reproduced next to an image of 'The Burning Ghat, Benares' (see illus. 76) whose inclusion derives largely from his emphasis on the religious and exotic in this picture of Indian life. Such scenes had long been used by missionary societies and others as evidence of the barbarism of non-Christian cultures. Edmund Evans's 1861 *Pictorial Missionary Map of the World,* for example, sold for a penny, included engraved scenes of the 'banks of the Ganges' and 'cruelty of hindoo idolatory' surrounding its world map of religions. India was largely coloured black signifying 'heathen'. Photographic lantern-slides, such as one of the 'Burning Ghat, Calcutta' (illus. 78), which shows the half-naked body of an Indian woman on the point of being burnt, were likely to have been similarly employed, along with stories of 'heathen' practices or even 'widow burning', to shock and thrill audiences at missionary society lectures and Sunday school classes. Indeed, this (originally coloured) lantern-slide is one of a set made by E. G. Wood of Cheapside, which consists mostly of Bible readings. Such photographs were also likely to have been posed, given the sensitive nature of the subject matter and the conspicuousness of a photographer in such settings. Fisher managed to avoid such negotiations by taking his photograph from a boat on the river at a distance.

For Fisher, the capturing of the culturally different was often the most enjoyable aspect of his duties on tour. Whereas his photographic surveys of buildings built under British authority elicited the barest of mention in his letters, his encounters with indigenous architecture and art frequently resulted in lengthy and enthusiastic comment.[64] Nevertheless, Fisher's frameworks for encountering difference had been laid down before he had even arrived in India, through both Mackinder's instructions and his own cultural assumptions. Fisher's experiences were certainly framed by European traditions of representation in which to travel to the East was to encounter the exotic and the sensual. As a painter himself, Fisher would have been even more likely to have known the works of British and French painters, such as Delacroix, which combined such elements in their representation of the Orient.

Both Fisher and Mackinder were drawing on European cultural traditions which produced the Orient as legible and knowable. It is thus certainly the case that both photographer and author followed the tradition of identifying 'Oriental' scenes or individuals held to be 'typical'. Mackinder, for example, referred to the city of Benares as 'an epitome of all Hindu India'. Commenting on Fisher's photograph of a man sitting on a bed of spikes there (illus. 79), Mackinder noted:

> In the narrow deep-shaded streets, and the sordid and tawdry purlieus of the temples may be seen many a typical scene of Eastern life. Here, for instance, close to Aurangzeb's Mosque, is a Fakir or religious enthusiast, to whom the alms of the faithful are due. He rests on this bed of spikes day and night.[65]

Fisher's photograph of 'A Fakir at Benares' thus represents for Mackinder a portrait of Eastern life typified as both ostentatious and ignoble. Fisher's view was not dissimilar, as he revealed in an explanatory note to Mackinder: 'Mahabi has been eight years like this – ghastly humbug – anyone could sit on these spikes without great discomfort, and I am told that Mahabi gets off after he has been lifted into a small tent every night.'[66]

This image is one of a number which, like the photograph of snake charmers which followed it in Mackinder's lectures and which is reproduced next to it in the textbook, were held to represent 'characteristic scenes' of Indian life. As I have noted, popular stereotypes of exotic difference gained particular currency from the ways photography was used as a form of classification in the creation and display of racial 'types'. Indeed, photographs of fakirs or snake charmers such as those made by the commercial photographers Shepherd and Robertson featured both in collections of British residents and tourists in India[67] and in photographic surveys of racial types such as *The People of India* (1868–75).

Mackinder's view of peoples of the Asian subcontinent and Fisher's choice of subjects for the camera were often informed by conventions of the racial 'type'. Many of Fisher's photographs thus follow a tradition of categorizing 'racial

Copyright.] [See page 59.

A FAKIR AT BENARES.

Copyright.] [See page 59.

SNAKE CHARMERS AT BENARES.

79 'A Fakir at Benares'; 'Snake Charmers at Benares', in H. J. Mackinder, *India: Eight Lectures* (1910).

types' through both their appearance and occupation. For example, Fisher noticed and photographed a blacksmith's workshop near Mandalay which, he said, 'betrayed the easy going way of Jack Burman'.[68] Not long after, he selected a 'Burman' for a portrait since he was 'a typical specimen of his race'.[69] Fisher's categorization of racial types was also framed by his imperial world-view. Fisher described the Burmese, for example, as easy to govern but prone to an 'indolence fatal to success', making them easy prey to foreign speculators. Fisher consequently advocated a strong imperial policy in Burma, securing trade advantages without allowing any 'Jap or Chinaman into our territory'.[70] Fisher's photographs were, in this respect, consistent with Mackinder's lectures where images of exotic 'others' function ultimately to distinguish the 'typically native' from the 'super-added' (and superior) European civilization.[71]

While Fisher certainly practised photography within general conventions of racial classification, the complexities of cultural difference could also disarm attempts at colonial categorization. Like many European travellers in the Indian subcontinent before him, Fisher was both fascinated and alarmed by bazaars.[72] However, he found the Mandalay Bazaar, for example, so complex that he could not 'catalogue' it with his camera, though he tried.[73] Similarly, he found a sculpture in Lucknow Museum so beautiful it was 'impossible to describe'.[74]

80 H. Fisher, 'Native Cutters Making Clothes for British Troops', 1908.

In his instructions to Fisher, Mackinder directed him to show and contrast India's 'old home industries' with the 'new industrial conditions in places like Bombay and Cawnpore'.[75] Fisher's photograph of 'Native Cutters Making Clothes for British Troops' (illus. 80), for instance, was made as part of his attempt to document Cawnpore's developing industries. With the help of this image, Mackinder described Indian economic development in evolutionary terms, explicitly comparing the changes in cities like Cawnpore with those of Britain during the Industrial Revolution. Borrowing from Fisher's letters, Mackinder described Cawnpore as 'the Manchester of India', claiming that here 'Western capital, Western ideas and Western organization are at work on a large scale'.[76] The quiet efficiency presented by the image of Indian men cutting out chalked outlines of clothes destined for British troops simultaneously reinforced the essential community of Empire. Even the khaki material used, Mackinder reported, was a product of both 'native labour and British machinery'.[77] However, Fisher himself was shocked by some of the working conditions and was aware of the selectivity of what he was shown and was allowed to photograph.[78] Finding traditional Indian trades far more to his liking, he was relieved to escape to the city of Agra where he could portray 'the picturesque side of old world India' and find, like many European travellers before him, a vision of an authentic, quasi-European past.[79]

By the turn of the twentieth century, railways were well-established colonial

81 H. Fisher, 'View of the Ruri end of the Lansdowne Bridge over the Indus - with train', 1908.

icons and Mackinder had little difficulty in using Fisher's photographs of trains, stations and dramatic sections of line, along with maps of the Indian Railway systems, as symbols of imperial progress and unity.[80] For example, on his travels on the North Western Railway between Lahore and Quetta, Fisher was given an official tour of the famous Lansdowne Bridge over the Indus (illus. 81).[81] This railway bridge, built by Robertson and Hecquet between 1887 and 1889, was a popular photographic target, images of it appearing in a number of contemporary publications describing Empire.[82] Using Fisher's photographs, Mackinder suggested it was 'perhaps the most remarkable bridge in India'. Pointing out the train on the bridge in Fisher's image, Mackinder noted that the view had been taken from Suttian, 'an old nunnery founded for women who preferred seclusion rather than the funeral pyre'.[83] Mackinder thus used this dramatic expanse of metal spanning the Indus not only as a sign of material strength but also as a metaphor for the moral improvement of India under the British. These representations of the Lansdowne railway bridge feature in Mackinder's final lecture (on the North-West Frontier) where the railways become the tendrils of British authority, forging through inhospitable territory to secure remote outposts such as Quetta and frontier stations such as Chaman.

Mackinder's lectures also delineated the strategic geography of particular regions. He was, for example, anxious that Fisher should document the

82 'Gurkhas at Bayonet Practice', in H. J. Mackinder, *India: Eight Lectures* (1910).

border areas of India such as the North-West Frontier.[84] Mackinder thus warned of 'Oriental despotism and race domination from the north-west' and images of strategic railways and military outposts were used to assuage such fears.[85] Scenes such as the Gurkha Rifles at bayonet practice (illus. 82) were similarly used to present an image of a secure and defensible Indian Empire. Yet grand displays of military power were not as numerous in the lantern-lectures on India as they were in contemporary commercial publications such as *The Queen's Empire*. The stress was rather upon the benign influence of imperial rule. 'The British', Mackinder argued, rather than simply conquering, 'have organized the peace and unity of India.'[86] It was, he asserted, the Courts of Justice that, 'more than any military power, betoken the essence of British rule in India'.[87]

Given that Fisher's photographs were planned to be shown in schools, it is perhaps unsurprising that these images also document different kinds of educational activities. Mackinder's lectures used such images to prove how 'Western civilization is permeating all India under British suzerainty.'[88] Fisher's images of schooling focused in particular on the display of orderly and disciplined activities, from wood-working classes at the Mayo School of Art in Lahore[89] to gymnastics instruction in the Government High School, Peshawar.[90] Using this image of disciplined schooling, Mackinder noted how even on the 'very borders of British rule, it is interesting to see the progress of Western education'.[91] There was a more general emphasis on paternalistic education throughout the lectures, in general reproducing the idea that Britain was guiding India to a mature level of civilization. This notion was advanced more explicitly in school geography textbooks such as W. G. Baker's *The British Empire* (1890), in which British India was described as 'a School for Asiatics, in which Europeans are the masters'.[92] In his own book, Fisher described Indians as 'foster-children' for whose education and welfare Britain was responsible.[93]

In all, Mackinder carefully arranged Fisher's photographs in his lectures to create a grand tour around the Indian Empire, attempting both to educate and to entertain his audience: to move them with images of 'Mutiny' monuments, excite them with pictures of exotic 'types' and impress them with displays of industrial progress. The COVIC's lectures on India employed a series of representations to propagate a vision of India's central place in the unity of Empire. In turn, they stressed that India's freedom was inconceivable outside the Empire. As one reviewer put it, the spirit of Mackinder's lectures was 'frankly imperialistic'.[94]

Projecting Empire

Following publication of the lantern-slide lectures on India, Mackinder's time was increasingly diverted from the COVIC by his many other commitments. Towards the end of 1910 he organized for a colleague, A. J. Sargent, Professor of Commerce at the University of London, to prepare the lantern-slide lectures on Canada, Australia and Imperial Outposts.[95] Nevertheless, these remained under Mackinder's overall supervision and follow closely the structure of his previous lectures, drawing on Fisher's photographs and operating firmly within the overall aims of the COVIC. Consequently, many of the themes that underlay the slides and lectures on India, far from being exclusively Orientalist modes of representation, recur in subsequent productions of the COVIC.

The COVIC's scheme in general emphasized a strategic view of imperial geography. In *The Sea Route to the East*, for example, Somaliland was described as having 'no meaning except in relation to the control of the sea'. Photographs of landscape were similarly used to convey the point of view of the commanding white man. Describing a visit to Sheikh, Somaliland, and accompanying a landscape view, Sargent's text notes: 'From the bungalow of the political officer we have a fine view down the long, steep pass, and can form some idea of the nature of the rugged country through which we are travelling.' Reading the landscape here was all the more important since, 'the geography of the country is all in favour of the native raider and against the civilized troops which attempt to catch him'. Photographs of a Somali warrior reinforced the message that 'the Somali is a born fighter, and his weapons are never very far away from him'.[96]

Photographs of different 'races' featured throughout the COVIC's lectures, yet their place was particularly circumscribed in the lectures on the 'settler colonies' of Canada and Australia. Since the photographer, Hugh Fisher, had made photographs of indigenous peoples in both Canada and Australia by visiting 'native reserves', the resulting lectures tended to confirm and confine them within such spaces. Australian Aborigines were thus represented as a primeval race doomed to gradual extinction, 'unless the native can change his character greatly'. Although the lectures were keen to stress improvements of indigenous peoples, they were also anxious to keep the 'natives' in their place. A picture of a 'Group of Natives, Queensland' was accompanied by the description: 'Here again, in Queensland, we see the native converted to European clothes, though he does not seem very comfortable in them.'[97] For both Canada and Australia, the lectures presented the decline of indigenous peoples as inevitable. They were certainly of little importance in comparison with the real value of the country. In discussing Australia, for example, Sargent declared:

> An area about equal to that of the United States could not be left in the sole occupation of a few thousand savages. Now, instead of the savage with his primitive tribal system, we have a white race, purely British in origin, with industry and agriculture of the most advanced type, and an elaborate political constitution of federated states.[98]

Similarly, the lectures on Canada included photographs of indigenous people which showed their 'primitive' customs and beliefs, as evidenced for instance by 'curious totem figures'. In short, the text asserted, 'The Indian hardly counts as a factor in Canadian development in the future; there are less than 150,000 in all, including the half-breeds, out of a population totalling nearly eight million.'[99] Depleted and outnumbered, the indigenous inhabitants were seen to pose no threat to the prosperity of the 'colonies of white settlement'.

The COVIC lectures on Canada and Australia were thus generally focused on prospects for settlement by British citizens. Scenes of agricultural and industrial development, from sheep-shearing in New South Wales to excavating a reservoir in Victoria, testify to the bright prospects of the Australian colonies. Slides of 'A settler's home' or 'A new homestead' in Queensland present an inviting view of prosperous, rural life. Such photographs have much in common with similar scenes employed by both the Canadian and the Australian governments in official lectures promoting emigration and with posters displaying scenes of life in 'Our Empire' (see illus. 73).

While the lectures were keen to show the benefits of emigration they were anxious not to overemphasize the place of 'foreigners' in the settler colonies. After a sequence of photographs showing settlements by Hungarians, Mennonites, Icelanders and Central Europeans, the text advised:

> But we must not imagine that Canada is becoming foreign. Most of the immigrants learn English and their descendants become Canadians. The predominant type is British . . . while most of the business of the country is in the hands of Canadians of British descent.[100]

Scenes of British Canadians fishing and lumbering, 'gathering maple sugar' or laying a section of the new Transcontinental Railway all testified to the industry and prosperity of the colony. One of Fisher's photographs, also reproduced in the COVIC textbook, pictured the prospects of British Columbia in the form of a giant tree (illus. 83). Fisher placed his hat on the side of this Douglas fir in Vancouver Park to emphasize its mammoth proportions. Just as Carleton Watkins's photographs of the sequoias in Yosemite in the early 1860s had confirmed the place of the 'Big Tree' in American national consciousness,[101] photographs of giant trees with smart Victorian children perched on roots or vehicles passing through a hollowed-out trunk became common icons of Britain's Empire in North America. Trees in general figured throughout the COVIC lectures on Canada, from scenes of 'A Sawmill' to 'Log Burling', as they were transformed from natural resources to trades, industries and products. However, the Douglas fir photographed by Fisher in Vancouver Park was preserved for posterity and for generations of visitors to pose for photographs beside it. Sargent's textbook described the Douglas firs in Vancouver Park as 'the last survivors of the forest',[102] for while the lectures celebrated the clearance of forest to build Vancouver and the Canadian Pacific Railway, they also lamented the loss of the trees and drew on photographs of surviving botanical giants to symbolize the beauty of the natural Canadian landscape.

A similar paradox is apparent in other scenes of colonial landscapes. Photographs of the Niagara Falls, for instance, draw on established conventions of romantic depiction to show the grandeur of this natural wonder. However, the text of the lecture notes, 'What appeals to the practical mind is not so much the beauty of the scene as the vast supply of power running to waste.'[103] The received message of the lectures thus depended upon the crucial conjunction between image and text.

On the whole, this marriage of imperial text with photograph was not difficult to achieve since Fisher usually photographed, as requested in his instructions, the overtly symbolic scenes. Thus colonial buildings and monuments figured prominently in the lectures as icons of imperial rule. Most lectures featured at least one picture of a statue of Queen Victoria, whose effigy could be found embellishing scores of imperial cities, from Bombay to Hong Kong. The lectures also prescribed standard ways of interpreting such

83 Hugh Fisher, 'Big tree (note hat on tree for scale)', 1909.

pictures of colonial art and architecture, even when initial appearances were not up to much. The British Residency in Penang, for example, might look like 'an insignificant hut' when seen alongside the palace of the Sultan of Selangor, 'but the hut represents efficiency in administration, while the Sikh sentry who guards it stands for the law and order which we have introduced'.[104] Scenes of law and order were prominently represented throughout the COVIC's lectures in slides of soldiers, sailors and policemen across the

84 Hugh Fisher, 'The kitchens of the same camp' (Christmas camp of Boy Scouts by Yarra River, Eltham, Victoria), 1909.

Empire, from 'Gurkhas at Bayonet Practice' to 'Fiji Native Constabulary'. Such images conveyed the idea that colonial order was maintained by duty and discipline, rather than sheer strength of numbers. One slide of the Royal North West Mounted Police at drill at Fort Saskatchewan was described thus: 'This fine body of men, only a few hundred in number, represent law and order throughout this vast territory, with its population of Indians and half-breeds, to its remotest corners.'[105]

Discipline and duty were at the core of the COVIC's message. After all, it was itself an argument for the Empire-wide benefits of schooling and citizenship. Schooling was taken as a sign of civilization and cultural advancement, as when the Christian inhabitants of the Friendly Islands were described as 'the most school-taught people in the Pacific'.[106] The spirit of imperial citizenship infused many of the lectures. Slides of scenes on the Yarra River outside Melbourne, for instance, showed the fun and games at the Christmas camp of the Boy Scouts (illus. 84) who, as the reader was reminded, 'are known in Australia as well as in our own part of the Empire'.[107] It is unlikely that the schoolboys receiving such visual instruction would have to be reminded of this fact. Baden-Powell's best-selling *Scouting for Boys* (1908) had been published only a few years earlier. Although the COVIC scheme was aimed at both girls and boys, the fact that *Scouting for Boys* should appear at the apex of the project is a striking reminder of the close connections between such schemes.

The Earl of Meath, Chairman of the COVIC, joined the Executive Committee of the Boy Scout movement in 1910 and played an active part in the movement until he was well into his eighties. Meath was involved in both organizations largely because they shared aims of fostering imperial citizenship in young people, particularly boys.

Evaluating the COVIC

On one level, the COVIC's scheme could be considered simply as an overly ambitious propaganda project. Although the COVIC managed to distribute many thousands of lantern-slides around the Empire, their complementary operations at home were plagued by protracted delays and lack of funds. Sales of the lantern-slide lectures in Britain were disappointing.[108] Most importantly, the COVIC was caught between its official responsibility and its populist ambitions. As part of the Colonial Office, it had to tread carefully with the information it propagated, yet it was obliged to enter a highly competitive field as a semi-commercial body. Though Mackinder continued to promote the COVIC's scheme and its mode of geographical visualization, he gradually excused himself from his practical duties at the COVIC. Although the possibility of transferring the scheme to a non-official imperial organization was considered in 1911, the COVIC did not want to lose either its commercial independence or its royal patronage. The scheme was kept alive after 1911 largely by the efforts of Charles Lucas and Everard im Thurn.[109] Through their efforts, photographs were collected, largely through correspondence with colonial governments,[110] for lantern-slide lectures on South Africa, the West Indies and the African Colonies.[111] With the onset of the First World War, however, the COVIC was discontinued, the completed parts of the scheme being transferred to the Royal Colonial Institute.

The significance of the work of the COVIC extends beyond its success or failure as a propaganda project. Although the scheme did not reach the mass British audience it had once aspired to, both its desire for a regenerated imperial spirit and its use of new means of visual instruction reflected much wider concerns within contemporary discourses of imperial education. In particular, the COVIC realized in institutional form Halford Mackinder's conception of an imperial geography based on the power of sight.[112] The 'visualizing power' described by Mackinder was part of the same spectrum of ideas on visual intelligence from which Baden-Powell drew his attempts to train the powers of observation of young boys.[113] Rudimentary 'geographical thinking', Mackinder argued, was apparent 'in the eye for the country which characterizes the fox hunter and the soldier'.[114] Mackinder's concern with 'geographical thinking', like Baden-Powell's idea of scouting, was derived in part from military training, yet was applied within a broader framework of civic duty, education and citizenship. Through geographical training, Mackinder believed, men could acquire 'the power of roaming at ease imaginatively over the vast surface of the globe'; the 'real geographer' was the man 'who *sees* the world-drama as he reads his morning paper'.[115] Mackinder's conception of imperial citizens was, again like Baden-Powell's, based largely on their being male. Moreover, 'thinking geographically' also involved adopting a particular perspective, a standpoint premised on British global authority. 'Through all,' Mackinder observed in 1911, 'let our teaching be from the British standpoint, so that finally we see the world as a theatre for British activity.'[116] The construction of the Empire as a vast visual display – the project of COVIC itself – was just such an exercise in imaginative geography.

The COVIC was not the only means used to project a geographical vision of the British Empire. Indeed, contemporary geography textbooks were invariably parallel exercises in imperial promotion. Mackinder himself produced a successful series of textbooks between 1906 and 1912 which made prolific use of both maps and photographs. As he noted in *Our Own Islands* (1906), an attempt to show children 'how so small a country can be the home of so great a people', 'No picture has been inserted for a merely decorative end, and the pupil should be led to question each view.'[117]Gradually building up the scope of the children's geographical imagination, *Our Own Islands* was followed by *Lands Beyond the Channel*, *Distant Lands* and, finally, *The Nations of the Modern World*.[118] The end point of this elementary geographical journey was reached when 'the pupil is asked to visualize with a single grasp our whole world of varied scene and incessant change'.[119] Such a global panorama was precisely what was sought by the COVIC.

The illustrative techniques and ideological perspective adopted in Mackinder's own textbooks, as well as those he wrote for the COVIC, were by no means unusual. Indeed, he took his cues from earlier popular illustrated publications such as Robert Brown's *The Countries of the World* (1876–81) which assume the structure of an imaginative 'grand tour', expounding a global geographical perspective in which Britain and her imperial interests were firmly centred.[120] Similarly, Blackie & Son's *Geographical Manuals* (1888–96) projected a vision of imperial geography through illustrative diagrams, views and maps.[121]

A number of contemporary textbooks employed even more novel photographic displays. W. Bisiker's *The British Empire* (1909), for example, reworked the traditional atlas, incorporating collages of photographs with maps, diagrams, statistics and descriptive text to provide a pictorial and cartographic survey of the world from a British imperial perspective. Offered as a contribution to debates on 'Education and Empire', the photographs and drawings and their explanations were 'included in the atlas with the idea of helping those, students especially, who have no intimate knowledge of the

85 'Southern Africa', W. Bisiker, *The British Empire* (1909).

Empire's features and resources, to associate the subjects illustrated with the colony'. Photographs and drawings showing the characteristic scenery, 'types of inhabitants', flora and fauna, numbered to correspond with a descriptive key, are here combined into a panorama of the country. The pictorial sheet of Southern Africa (illus. 85), for example, with its images of the African lion and Victoria Falls, was further framed by maps of political boundaries; animal, vegetable and mineral products; climate; relief; population density; and statistical information on the government, history and finance of the territories. Bisiker intended them to be read together as 'an epitome of the geography' of a colonial area.[122]

This technique of photographic collage was adopted by a number of textbooks of the time. *Philips' Picture Atlas of the British Empire* (1911), for example, contained similar 'Empire Pictures' consisting of organic photographic compositions juxtaposed with graphic information on the trade, defence, communications, population and territorial control of the Empire. Comparative statistical information on the 'home and colonial' areas and population of the world's empires was thus presented using scale outlines of countries and pictures of 'racial types' to convey in graphic efficiency the superiority of Britain's Empire. The pictorial design of these Empire pictures owed much to trends within education. Yet they also drew on the contemporary

7 T. Baines, storekeeper's notebook (photocopy), Brenthurst Library, MS 029, pp. 54, 114.
8 H. C. Rawlinson, 'Presidential Address', *JRGS*, XLV (1875), p. ccx.
9 D. Livingstone to N. Bedingfeld, 10 April 1858, in *The Zambesi Expedition of David Livingstone 1858–1863: The Journal Continued with Letters and Dispatches Therefrom*, ed. J. P. R. Wallis (London, 1956), II, p. 423. A similar summary of the official aims of the expedition is given in D. Livingstone and C. Livingstone, *Narrative of an Expedition to the Zambesi and its Tributaries and of the Discovery of the Lakes Shirwa & Nyassa, 1858–1864* (London, 1865), p. 9.
10 D. Livingstone to C. Livingstone, 10 May 1858, in *Zambesi Expedition*, ed. Wallis, p. 431.
11 Charles Livingstone to Austin Layard, n.d., FO97/322, f.156, cited in G. W. Clendennen, *Charles Livingstone: A Biographical Study with Emphasis on his Accomplishments on the Zambesi Expedition, 1858–1863* (Ph.D. thesis University of Edinburgh, 1978), p. 255.
12 Richard Owen to Charles Spring-Rice, 9 December 1863, FO97/322, f.158, cited ibid., p. 256.
13 Livingstone and Livingstone, *Narrative of an Expedition to the Zambesi*, p. vii.
14 See R. Coupland, *Kirk on the Zambesi: A Chapter of African History* (Oxford, 1928), pp. 54, 129. See also A. D. Bensusan, *Silver Images: The History of Photography in Africa* (Cape Town, 1966), pp. 24–5.
15 A total of twenty-nine of Kirk's Zambezi photographs, as paper and glass negatives and prints, survive in a private collection on loan to the NLS, Acc. 9942/40 and 41.
16 9 July 1858, cited in *Baines on the Zambesi, 1858–1859*, ed. E. C. Tabler, E Axelson and E. N. Katz (Johannesburg, 1982), p. 120.
17 Journal/letter, Charles Livingstone to Hariette Livingstone, 14 September–21 December 1858, G5/10 National Museum, Livingstone, Zambia, cited in Clendennen, *Charles Livingstone*, p. 251.
18 Ibid.
19 Ibid.
20 T. Barringer, 'Fabricating Africa: Livingstone and the Visual Image 1850–1874' in *David Livingstone and the Victorian Encounter with Africa*, exhibition catalogue, ed. J. M. MacKenzie: National Portrait Gallery, London (London, 1966), pp. 169–200.
21 D. Livingstone to T. Baines, 18 May, 1858, in *Zambesi Expedition*, ed. Wallis, p. 434.
22 W. Allen, *Picturesque Views on the River Niger. Sketched During Lander's Last Visit in 1832–33* (London, 1840), preface.
23 D. Livingstone, 'Extracts from the Despatches of Dr David Livingstone to the Right Honourable Lord Malmesbury, *JRGS*, XXXI (1861), pp. 256–96.
24 D. Livingstone to Lord Malmesbury, 17 December 1858, in *Zambesi Expedition*, ed. Wallis, p. 294.
25 Ibid., p. 299.
26 T. Jeal, *Livingstone* (New York, 1973), pp. 202–14. See also Coupland, *Kirk on the Zambezi*, p. 136.
27 J. Kirk, 'Kebrabasa Rapids, from a sketch by Dr Kirk, Morumbwa cataract' (*c.* 1858); 'The "Ma Robert" on the Zambesi at Lupata' (*c.* 1859); 'Lupata Gorge, from right bank looking down. July 1859', NLS Archives, Acc. 9942/40.
28 J. Kirk, 'Murchison Rapids, River Shire' (*c.* 1859), NLS Archives, Acc. 9942/40.
29 R. A. Stafford, *Scientist of Empire: Sir Roderick Murchison, Scientific Exploration and Victorian Imperialism* (Cambridge, 1989), pp. 28–9.
30 Livingstone and Livingstone, *Narrative of an Expedition to the Zambesi*, p. 6.
31 P. D. Curtin, *The Image of Africa: British Ideas and Action, 1780–1850* (London, 1965); Brantlinger, 'Victorians and Africans', pp. 166–203.
32 Curtin, *Image of Africa*, pp. 58–87.
33 Earl Grey, 'Address to the Royal Geographical Society', *PRGS*, IV (1860), pp. 117–209.
34 J. Kirk, 'The Extent to Which Tropical Africa is Suited for Development by the White Races, or Under Their Superintendence', *Report of the Sixth International Geographical Congress* (London, 1896), p. 526. See also Curtin, *Image of Africa*, pp. 185–7, 349–53;

D. N. Livingstone, 'The Moral Discourse of Climate: Historical Considerations on Race, Place and Virtue', *JHG*, XVII (1991), pp. 413–34.

35 Anon., 'The Application of the Talbotype', *Art Union* (1846), p. 195.

36 For an example of this procedure in travel writing, see M. L. Pratt, *Imperial Eyes: Travel Writing and Transculturation* (London, 1992), p. 61.

37 F. Galton, 'Zanzibar', *The Mission Field*, VI (1861), p. 128.

38 Earl Grey, 'Address to the Royal Geographical Society', pp. 117–209.

39 P. J. Marshall and G. Williams, *The Great Map of Mankind: British Perceptions of the World in the Age of Enlightenment* (London, 1982), pp. 227–57.

40 Curtin, *Image of Africa*, pp. 61–2.

41 A similar photograph by Kirk with the same title, 'Lupata July 13th 1859', is captioned, 'Vegetation at Lupata, Baobab stem in front, with soft twiners, and the poison acid juiced Tree Euphorbia which touched up Rowe the second engineer in this neighbourhood last month', NLS Archives Acc. 9942/40.

42 J. Chapman, *Travels in the Interior of South Africa 1849–1863: Hunting and Trading Journeys from Natal to Walvis Bay & Visits to Lake Ngami & Victoria Falls*, ed. E. C. Tabler (Cape Town, 1971).

43 J. Chapman, 'Notes on South Africa', *PRGS*, V (1860), pp. 17–18.

44 See 'Extracts from a Letter from J. Kirk', *JRGS*, XXXIV (1865), pp. 290–92.

45 J. Thomson, 'Exploration with the Camera', *BJP*, XXXII (1885), p. 373.

46 See T. J. Last, 'On the Society's Expedition to the Namuli Hills, East Africa', *PRGS*, IX (1887), pp. 467–78; T. J. Last, 'A Journey from Blantyre to Angoni-land and Back', *PRGS*, IX (1887), pp. 177–87.

47 *Photographic Views of Blantyre, BCA* (1900/01–1905), FCOL Photos, Malawi 1. See, for example, pl. 7: 'Blantyre Mission Church'; pl. 21: 'Coffee Plantations BCA. Kirk Mountains in Distance'; pl. 40: 'Croquet Lawn near Vice-Consulate'.

48 Ibid. See, for example, pl. 97: 'Through the Tangled Forest'; pl. 96: 'Rocks and Roots'.

2 *Framing the View*

1 R. Herchkowitz, *The British Photographer Abroad: The First Thirty Years* (London, 1980); A. Rouillé, 'Exploring the World by Photography in the Nineteenth Century' in *A History of Photography: Social and Cultural Perspectives*, ed. A. Rouillé and J. Lemagny (Cambridge, 1987), pp. 53–9; R. Fabian and H. Adam, *Masters of Early Travel Photography* (London, 1983).

2 N. M. P. Lerebours, *Excursions daguerriennes: representant les vues et les monuments anciens et modernes les plus remarquables du globe* (Paris, 1841–4).

3 E. Onne, *Photographic Heritage of the Holy Land 1839–1914* (Manchester, 1980); *Images of the Orient: Photography and Tourism 1860–1900*, exhibition catalogue, ed. P. Faber, A. Groeneveld and H. Reedijk: Museum voor Volkenkunde, Rotterdam (Amsterdam and Rotterdam, 1986).

4 J. Talbot, *Francis Frith* (London, 1985).

5 F. Galton to Dr Norton Shaw, 25 March 1859 and 2 May 1859, RGS Archives.

6 R. J. Fowler, 'Letter from Paris, April 29 1867', *BJP*, XIV (1867), pp. 212–13.

7 See J. M. MacKenzie, 'Introduction' in *Popular Imperialism and the Military 1850–1950*, ed. J. M. MacKenzie (Manchester, 1992), p. 5.

8 D. Cosgrove, *Social Formation and Symbolic Landscape* (London, 1984); D. Cosgrove, 'Prospect, Perspective and the Evolution of the Landscape Idea', *TIBG*, X (1985), pp. 45–62; D. Cosgrove and S. Daniels, eds, *The Iconography of Landscape: Essays on the Symbolic Representation, Design and Use of Past Environments* (Cambridge, 1988); S. Daniels, *Fields of Vision: Landscape Imagery and National Identity in England and the United States* (Cambridge, 1993); S. Schama, *Landscape and Memory* (London, 1995).

9 W. J. T. Mitchell, 'Imperial Landscape' in *Landscape and Power*, ed. W. J. T. Mitchell (Chicago, 1994), p. 17.

10 B. Smith, *European Vision and the South Pacific* (London, 1985). See also J. Hackforth-Jones, 'Imaging Australia and the South Pacific' in *Mapping the Landscape*, ed. N. Alfrey and S. Daniels (Nottingham, 1990), pp. 13–17.

11 G. Batchen, 'Desiring Production Itself: Notes on the Invention of Photography' in *Cartographies: Poststructuralism and the Mapping of Bodies and Spaces*, ed. R. Diprose and R. Ferrell (London, 1991), pp. 13–26.

12 R. Krauss, 'Photography's Discursive Spaces: Landscape/View', *Art Journal*, XLII (1982), pp. 311–20.

13 A. Birrell, 'Survey Photography in British Columbia, 1858–1900', *BC Studies*, LII (1982), pp. 39–60; M. T. Hadley, 'Photography, Tourism and the CPR: Western Canada, 1884–1914' in *Essays on the Historical Geography of the Canadian West: Regional Perspectives on the Settlement Process*, ed. L. A. Rosenvall and S. M. Evans (Calgary, 1987), pp. 48–69.

14 J. M. Schwartz, 'Images of Early British Columbia: Landscape Photography, 1858–1888' (M.A. thesis, Department of Geography, University of British Columbia, 1977).

15 D. Mattison and D. Savard, 'The North-west Pacific Coast: Photographic Voyages 1866–81', *History of Photography*, XVI (1992), pp. 268–88.

16 Krauss, 'Photography's Discursive Spaces'.

17 For a similar perspective, see E. Jussim and E. Lindquist-Cock, *Landscape as Photograph* (New Haven, 1985).

18 Smith, *European Vision*, pp. 56–8, 106–7.

19 See also J. Snyder, 'Territorial Photography' in *Landscape and Power*, ed. Mitchell, pp. 175–201.

20 S. Bourne, 'Photography in the East', *BJP*, X (1863), p. 268.

21 Ibid.

22 S. Bourne, 'Ten Weeks with the Camera in the Himalayas', *BJP*, XI (1864); 'Narrative of a Photographic Trip to Kashmir (Cashmere) and Adjacent Districts', *BJP*, XIII–XIV (1866–7); 'A Photographic Journey through the Higher Himalayas', *BJP*, XVI–XVII (1870–71).

23 A. Scharf, *Pioneers of Photography: An Album of Pictures and Words* (New York, 1976), pp. 87–102; A. Ollman, *Samuel Bourne: Images of India* (California, 1983); A. Ollman, 'Samuel Bourne: The Himalayan Images 1863–69', *Creative Camera* (1983), pp. 1,122–9.

24 See, however, C. Pinney, 'Imperial Styles of Photography: Some Evidence from India' in *Photographs as Sources for African History*, ed. A. Roberts (1988), pp. 20–27; G. D. Sampson, 'The Success of Samuel Bourne in India', *History of Photography*, XVI (1992), pp. 336–47.

25 Anon., 'Letters', *BJP*, XVI (1869), p. 477.

26 Anon., 'Review', ibid., p. 571.

27 Fowler, 'Letter from Paris, April 29 1867', pp. 212–13.

28 M. Archer, *British Drawings in the India Office Library* (London, 1969), II, pp. 574–99. See also G. H. R. Tillotson, 'The Indian Picturesque: Images of India in British Landscape Painting, 1780–1880' in *The Raj: India and the British 1600–1947*, exhibition catalogue, ed. C. A. Bayly: National Portrait Gallery, London (London, 1990), pp. 141–51.

29 Bourne, 'Narrative of a Photographic Trip to Kashmir'.

30 J. Murray, *Picturesque Views in the North-Western Provinces of India. Photographed by J. Murray, with Descriptive Letter-Press by J. T. Boileau* (London, 1859).

31 M. Clarke, *From Simla Through Ladac and Cashmere, 1861* (Calcutta, 1862).

32 Ibid., pl. 1: 'Simla'; pl. 5: 'A Saungur, or Hill Bridge, Over the River Sutlej, at Képoo, Below Kotegōōr'; pl. 25: 'The Visitor's Reach, Sreenuggur'.

33 Ibid., pl. 27: 'The Shalimar Gardens, Sreenuggur'; pl. 30: 'The 2nd or Ameera Bridge, Sreenuggur'; Bourne, 'Narrative of a Photographic Trip to Kashmir', pp. 4–5.

34 Clarke, *From Simla Through Ladac and Cashmere, 1861*, pl. 21: 'An Old Imperial Bridge Near Sreenuggur, Cashmere'.

35 Bourne, 'Narrative of a Photographic Trip to Kashmir', p. 5.

36 Bourne, 'Photography in the East', p. 345.

37 Bourne, 'Narrative of a Photographic Trip to Kashmir', p. 619.

38 Bourne, 'Photography in the East', p. 346.
39 Ibid.
40 Pinney, 'Imperial Styles of Photography'.
41 Bourne, 'Narrative of a Photographic Trip to Kashmir', p. 474.
42 Anon., 'Indian Photographs. By S. Bourne', *BJP*, XIV (1867), p. 17.
43 Bourne, 'Narrative of Photographic Trip to Kashmir', p. 474.
44 Ibid., pp. 474, 499; Bourne, 'A Photographic Journey', p. 603.
45 Bourne, 'Narrative of a Photographic Trip to Kashmir', p. 39.
46 Ibid., pp. 4–5.
47 Ibid., p. 39.
48 M. Alloula, *The Colonial Harem* (Manchester, 1987); Sarah Graham-Brown, *Images of Women: The Portrayal of Women in Photography of the Middle East, 1860–1950* (London, 1988).
49 E. W. Said, *Orientalism* (London, 1978); J. M. MacKenzie, *Orientalism: History, Theory and the Arts* (Manchester, 1995); R. Lewis, *Gendering Orientalism: Race, Feminity and Representation* (London, 1996).
50 N. N. Perez, *Focus East: Early Photography in the Near East (1839–1885)* (New York, 1988), p. 105.
51 C. Pinney, 'Classification and Fantasy in the Photographic Construction of Caste and Tribe', *Visual Anthropology*, 3 (1990), pp. 259–88.
52 Bourne, 'A Photographic Journey', p. 570.
53 Ibid.
54 Bourne, 'Ten Weeks with the Camera', p. 50.
55 P. H. Egerton, *Journal of a Tour through Spiti, to the Frontier of Chinese Thibet, with Photographic Illustrations* (London, 1864), preface.
56 Ibid.
57 Sampson, 'Success of Samuel Bourne', p. 339.
58 Egerton, *Journal*, p. 8.
59 Bourne, 'Ten Weeks with the Camera', p. 70.
60 Egerton, *Journal*, facing p. 60.
61 Ibid., pl. 12: 'The Hamta Pass'.
62 See also Bourne, 'A Photographic Journey', pp. 613–14.
63 Egerton, *Journal*, p. 10.
64 Ibid., pl. 14: 'Shigri Glacier – Upper View'; pl. 15: 'Rock on a Pedestal of Ice'; pl. 16: 'Shigri Glacier – Lower View'; pl. 17: 'Shigri Glacier – From the River'.
65 Bourne, 'A Photographic Journey', p. 629.
66 Ibid.
67 Egerton, *Journal*, p. 12.
68 Ibid., p. 7.
69 Bourne, 'A Photographic Journey', p. 150.
70 Scharf, *Pioneers of Photography*, pp. 87–102.
71 Sampson, 'Success of Samuel Bourne'.
72 Cosgrove, 'Prospect, Perspective and the Evolution of the Landscape Idea', p. 55.
73 Bourne, 'Ten Weeks with the Camera', p. 51; Bourne, 'Narrative of a Photographic Trip to Kashmir', pp. 559–60.
74 M. Andrews, *The Search for the Picturesque Landscape: Aesthetics and Tourism in Britain, 1760–1800* (Standford, 1989).
75 P. Hansen, 'Albert Smith, the Alpine Club, and the Invention of Mountaineering in Mid-Victorian Britain', *Journal of British Studies*, XXXIV (1995), pp. 300–324.
76 P. Hansen, 'Vertical Boundaries, National Identities: British Mountaineering on the Frontiers of Europe and the Empire, 1868–1914', *Journal of Imperial and Commonwealth History*, XXIV (1996), pp. 48–71.
77 Bourne, 'A Photographic Journey', p. 16.
78 Ibid., pp. 39–40.

79 Ibid., pp. 98–9.
80 Bourne, 'Ten Weeks with the Camera', p. 51.
81 J. Ruskin, *Modern Painters* (Boston, 1875), IV, pp. 133–4, cited in Schama, *Landscape and Memory*, p. 513.
82 Ibid., pp. 506–13.
83 T. Hoffman, 'Exploration in Sikkim: To the North-East of Kanchinjinga', *PRGS*, XIV (1892), pp. 613–18.
84 Anon., 'Photographs: Twenty-three Photographs of Mountain Scenery in Sikkim', *PRGS*, XV (1893), p. 288. See RGS photos PR/031591–677.
85 J. Thomson, *Illustrations of China and Its People, a Series of Two Hundred Photographs with Letterpress Description of the Places and People Represented* (London, 1873–4), I, introduction.
86 Ibid., 'From Hankow to the Wu-Shan Gorge, Upper Yangtsze', III, pls 17–24.
87 Ibid., I, introduction.
88 Ibid., 'The Tsing-Tan Rapid, Upper Yangtsze', III, pl. 22, no. 49.
89 J. Thomson, 'Photography and Exploration', *PRGS*, n.s. XIII (1891), p. 672.
90 A. S. Bickmore, 'Sketch of a Journey Through the Interior of China', *PRGS*, XII (1867), pp. 51–4.
91 A. Cotton, 'On a Communication between India and China', *JRGS*, XXXVII (1867), pp. 231–9.
92 J. Thomson, 'The Gorges and Rapids of the Upper Yangtsze', *Report of the BAAS*, I (1874), pp. 86–7.
93 Anon., 'Illustrations of China and Its People', *BJP*, XX (1873), p. 570.
94 G. Tissandier, *A History and Handbook of Photography* (London, 1876), p. 14.
95 S. White, *John Thomson: Life and Photographs – The Orient, Street Life in London, Through Cyprus with the Camera* (London, 1985), p. 30.
96 J. Thomson, *The Straits of Malacca, Indo-China and China or Ten Years' Travels, Adventures and Residence Abroad* (London, 1875); J. Thomson, *Through China with a Camera* (London, 1898); R. Brown, *The Countries of the World: Being a Popular Description of the Various Continents, Islands, Rivers, Seas, and Peoples of the Globe* (London, 1876–81).
97 See J. Thomson to H. W. Bates, 12 June 1875, RGS Archives; *Congrès internationale des sciences géographique* (Paris, 1876), II, p. 432.
98 W. Allen, *Picturesque Views on the River Niger, Sketched during Lander's Last Visit in 1832–33* (London, 1840), preface.
99 Thomson, 'Silver Island, River Yangtsze', *Illustrations of China*, III, pl. 8, no. 16.
100 Ibid., 'Canton', I, pl. 16.
101 Ibid., 'Shanghi Bund in 1869', III, pl. 4l, no. 7.
102 Ibid.
103 Ibid., 'Part of Shanghi Bund in 1872', III, pl. 5, no. 8.
104 Ibid., introduction.
105 Ibid.
106 Ibid., 'Hong-kong', I, pl. 2.
107 Ibid., introduction.
108 Said, *Orientalism*, pp. 204–5.
109 Alloula, *The Colonial Harem*; Perez, *Focus East*.
110 See, for example, Thomson, 'Cenotaph Erected to the Banjin Lama of Thibet', *Illustrations of China*, IV, pl. 15, and 'The Open Altar of Heaven and the Temple of Heaven, Pekin', IV, pl. 26.
111 Ibid., 'The Great Wall of China', IV, pl. 24, no. 56.
112 J. Thomson, *Through Cyprus with the Camera, in the Autumn of 1878* (London, 1879), 2 vols.
113 Ibid., I, preface.
114 Ibid., 'The Sea Shore, Larnaca', I, pl. 3. For other comments on improving marshlands see ibid., 'Famagosta', I, preface; II, pl. 45.

115 See, for example, ibid., 'The Pines of Olympus (Troodos)', I, pl. 35.
116 See, for example, ibid., 'Kerynia Harbour', I, pl. 21; 'Famagosta Harbour', II, pl. 46.
117 Ibid., 'The Front of St Katherine's Church (Now a Mosque) Famagosta', II, pl. 48.
118 Ibid., 'Ruins at Famagosta', II, pl. 47.
119 Ibid., 'Entrance to the Cathedral, Famagosta', II, pl. 49.
120 Ibid., 'St Nicholas, Nicosia', II, pl. 15.
121 J. W. Lindt, *Picturesque New Guinea* (London, 1887), pp. vii–viii. A five-volume set of numbered photographs, mounted on card with captions only, with the same title, was also produced: J. W. Lindt, *Picturesque New Guinea*, FCOL, Papua New Guinea, 2.
122 Lindt, *Picturesque New Guinea*, p. viii.
123 Ibid., p. 28.
124 Lindt, 'Reach of Laloki River, near Badeba Creek, *Picturesque New Guinea*, FCOL, pl. 31.
125 See R. Holden, *Photography in Colonial Australia: The Mechanical Eye and the Illustrated Book* (Sydney, 1988), p. 31.
126 Lindt, *Picturesque New Guinea*, p. 30.
127 F. F. Statham, 'On the Application of Photography to Scientific Pursuits', *BJP*, VI (1860), pp. 192–3.
128 S. Sontag, *On Photography* (London, 1978), p. 7.

3 *The Art of Campaigning*

1 F. F. Statham, 'On the Application of Photography to Scientific Pursuits', *BJP*, VI (1860), p. 193.
2 Anon., 'Photography Applied to the Purposes of War', *Art Journal*, VI (1854), p. 152.
3 S. Highley, 'On the Means of Applying Photography to War Purposes in the Army and Navy', *Report of the BAAS*, XXIV (1854), p. 70.
4 C. E. Callwell, *Small Wars: Their Principles and Purpose* (London, 1899), pp. 1, 5.
5 See R. H. MacDonald, 'A Poetics of War: Militarist Discourse in the British Empire, 1850–1918', *Mosaic*, XXIII, 3 (1990), pp. 17–36; J. Springhall ' "Up Guards and At Them!": British Imperialism and Popular Art, 1880–1914' in *Imperialism and Popular Culture*, ed. J. M. MacKenzie (Manchester, 1986), pp. 49–72.
6 J. M. MacKenzie, ed., *Popular Imperialism and the Military 1850–1950* (Manchester, 1992).
7 For accounts of the campaign, see A. Moorehead, *The Blue Nile* (London, 1962), pp. 211–80; D. Bates, *The Abyssinian Difficulty: The Emperor Theodorus and the Magdala Campaign 1867–68* (Oxford, 1979).
8 Callwell, *Small Wars*, p. 6.
9 See G. A. Henty, *The March to Magdala: Letters Reprinted from the "Standard" Newspaper* (London, 1868); H. M. Stanley, *Coomassie and Magdala: The Story of Two British Campaigns in Africa* (London, 1874).
10 Bates, *Abyssinian Difficulty*, p. 130.
11 F. Harcourt, 'Disraeli's Imperialism, 1866–1868: A Question of Timing', *History Journal*, XXIII (1980), pp. 87–109.
12 For a complete album, see *Abyssinian Expedition*, FCOL Ethiopia/1. See also RGS PR/036171–036246; NAM Photos 7604–43. For some views not reproduced in the official album, see NAM Photos 6510–222 (2–10).
13 T. J. Holland and H. M. Hozier, *Record of the Expedition to Abyssinia, Compiled by Order of the Secretary of State for War* (London, 1870).
14 Royal Engineers, 'Released Prisoners (Europeans)', RGS PR/036177.
15 S. Bourne, 'Photography in the East', *BJP*, X (1863), p. 268.
16 Bourne and Shepherd, *2nd Afghan War, 1878–80*, NAM Photos, 5504–42.
17 J. Burke, *Afghan War 1878–79, Peshawur Valley Field Force, Album of 98 Photographs*, IOL Photo 430/2. See also R. Desmond, *Victorian India in Focus: A Selection of Early*

Photographs from the Collection in the India Office Library and Records (London, 1982), p. 65.

18 H. Gernsheim and A. Gernsheim, *Roger Fenton: Photographer of the Crimean War* (London, 1954).

19 See Desmond, *Victorian India in Focus*, pl. 52, p. 68.

20 J. McCosh, *Album of 310 photographs* (1848–53), NAM Photos 6204–3. Compiled by McCosh himself in the late 1850s, this album is the largest known surviving collection of his work.

21 J. McCosh, *Advice to Officers in India* (London, 1856), cited in R. McKenzie, ' "The Laboratory of Mankind": John McCosh and the Beginnings of Photography in British India', *History of Photography*, XI (1987), p. 109.

22 Ibid., p. 114.

23 J. McCosh, *Topography of Assam* (Calcutta, 1837).

24 See J. McCosh, 'An Account of the Mountain Tribes in the Extreme NE Frontier of Bengal', *Journal of the Asiatic Society of Bengal*, V (1836), pp. 193–208.

25 Dr M'Cosh [*sic*], 'On the Various Lines of Overland Communication between India and China', *PRGS*, IV (1860), pp. 47–54.

26 See, for example, R. W. Porter, *History of the Corps of Royal Engineers* (Chatham, 1889) II, pp. 187–8.

27 J. Falconer, 'Photography and the Royal Engineers', *Photographic Collector*, II (1981), pp. 33–64.

28 Porter, *History of the Corps of Royal Engineers*, I, p. 4.

29 Established in 1812, the Royal Engineer's Establishment at Chatham was renamed the School of Military Engineering in 1869. See Ibid., II, pp. 183–4.

30 J. Donnelly, 'On Photography and Its Application to Military Purposes', *BJP*, VII (1860), pp. 178–9.

31 Ibid., p. 179.

32 J. Spiller, 'Photography in Its Application to Military Purposes', *BJP*, X (1863), p. 486.

33 Anon., 'The Application of Photography to Military Purposes', *Nature*, II (1870), pp. 236–7.

34 H. B. Pritchard, 'Photography in Connection with the Abyssinian Expedition', *BJP*, XV (1868), pp. 601–3.

35 Porter, *History of the Corps of Royal Engineers*, II, p. 6.

36 R. Hyde, *Panoramania! The Art and Entertainment of the 'All-Embracing' View* (London, 1988), pp. 179–98.

37 See 'discussion after Markham's paper', *PRGS*, XII (1868), p. 116.

38 Porter, *History of the Corps of Royal Engineers*, I, p. 5; II, p. 1.

39 H. St Clair Wilkins, *Reconnoitring in Abyssinia: A Narrative of the Proceedings of the Reconnoitring Party, Prior to the Arrival of the Main Body of the Expeditionary Field Force* (London, 1870), p. 312.

40 Pritchard, 'Photography in Connection with the Abyssinian Expedition', p. 603.

41 W. Abney, *Instruction in Photography: For Use at the SME Chatham* (Chatham, 1871), p. 1. Abney's later version, *Instruction in Photography* (London, 1874), ran to some eleven editions.

42 Pritchard, 'Photography in Connection with the Abyssinian Expedition', p. 602.

43 Ibid., p. 603.

44 D. R. Stoddart, 'The RGS and the "New Geography": Changing Aims and Changing Roles in Nineteenth Century Science', *GJ*, CXLVI (1980), pp. 190–202; D. R. Stoddart, 'Geography and War: The "New Geography" and the "New Army" in England, 1899–1914', *Political Geography*, XI (1992), p. 89.

45 C. W. Wilson, 'Address to the Geographical Section of the British Association', *PRGS*, XIX (1874), p. 63.

46 Ibid., p. 65.

47 C. W. Wilson to D. Freshfield, 6 March 1892, RGS Archives.

48 B. Becker, *Scientific London* (London, 1874), cited in F. Driver, 'Geography's Empire: Histories of Geographical Knowledge', *Environment and Planning D: Society and Space*, X (1992), p. 29.
49 See 'Accessions to the Library since the Last Meeting', *PRGS*, XII (1867), pp. 1–4; 'Report of the Council', *JRGS*, XXXVIII (1868), p. viii.
50 A. C. Cooke, *Routes in Abyssinia* (London, 1867).
51 See RGS Prints D108/023970–023981.
52 W. Allen, *Picturesque Views on the River Niger, Sketched during Lander's Last Visit in 1832–33* (London, 1840).
53 M. O'Reilly, 'Twelve Views in the Black Sea and the Bosphorus' (1856), RGS Prints D108, I, 17–30.
54 J. Ferguson (Lithographer), *Views in Abyssinia* (1867), RGS Prints D108/125–136.
55 See 'discussion after Markham's paper', *PRGS*, XII (1868), p. 118.
56 H. James and R. I. Murchison, *Ordnance Survey: Report of the Committee on the Reduction of the Ordnance Plans by Photography* (London, 1859).
57 F. Galton, 'On Stereoscopic Maps, Taken from Models of Mountainous Countries', *JRGS*, XXXV (1865), p. 101.
58 Cited in B. Parritt, *The Intelligencers: The Story of British Military Intelligence up to 1914* (Ashford, 1983), p. 97.
59 See R. I. Murchison, 'Anniversary Address', *JRGS*, XIV (1844), pp. cvi–cxx, cited in R. Stafford, *Scientist of Empire: Sir Roderick Murchison, Scientific Exploration and Victorian Imperialism* (Cambridge, 1989), p. 154.
60 R. I. Murchison, 'Presidential Address', *PRGS*, XII (1867), p. 6.
61 C. R. Markham, 'Geographical Results of the Abyssinian Expedition', *JRGS*, XXXVIII (1868), pp. 12–49; also in *PRGS*, XII (1868), pp. 113–19, 298–301. See also C. R. Markham, *A History of the Abyssinian Expedition* (London, 1869).
62 See 'discussion after Markham's paper', *PRGS*, XII (1868), p. 301.
63 E. W. Said, *Orientalism* (London, 1978), pp. 42–3, 80–93.
64 MacKenzie, ed., *Popular Imperialism and the Military 1850–1950*, pp. 1–24.
65 Markham, *History of the Abyssinian Expedition*, p. 389.
66 Ibid., p. 236.
67 See, for example, Said, *Orientalism*, pp. 84–6.
68 See 'discussion after Markham's paper', *PRGS*, XII (1868), pp. 115–19.
69 RGS Museum 63/120.1; 94/120.1. Theodore's crown was returned to Ethiopia in 1925.
70 Hyde, *Panoramania!*, pp. 169–78.
71 Bates, *Abyssinian Difficulty*, p. 214.
72 H. Schaw, 'Notes on Photography', *Professional Papers of the Corps of Royal Engineers*, n.s. IX (1860), pp. 108–28.
73 See A. Birrell, 'Classic Survey Photographs of the Early West', *Canadian Geographical Journal*, XCI, (1975), pp. 12–19; A. Birrell, 'Survey Photography in British Columbia, 1858–1900', *BC Studies*, LII (1982), pp. 39–60.
74 C. Wilson, 'Report on the Indian Tribes Inhabiting the Country in the Vicinity of the 49th Parallel of North Latitude', *TESL*, n.s. IV (1865), pp. 275–332.
75 N. Alfrey and S. Daniels, eds, *Mapping the Landscape: Essays on Art and Cartography* (Nottingham, 1990).
76 Allen, 'The Confluence of the Rivers Niger and Chadda' in *Picturesque Views on the River Niger*, pl. 13.
77 R. Baigre, ' "Deema" 3rd Halting Ground up the Tekoonda Pass', watercolour sketch, November 1867, photographed by Captain Sellon, RE, NAM Photos 6510–222, pl. 37, reverse.
78 Anon., 'Panoramic View of Plateau', n.d., two-photograph panorama annotated in black ink, NAM Photos 6510–222, pl. 21.
79 See R. Baigre, 'Action of Arogee, Under the Heights of Fahla and Selassie Fought 10 April 1868', lithograph by J. Ferguson, Topographical Department of the War Office, RGS

Prints X441/022937; Lt.-Colonel R. Baigre, 'The Capture of Magdala, 13 April 1868', lithograph by J. Ferguson, Topographical Department of the War Office, RGS Prints X441/022938. Both these views were reproduced in Holland and Hozier, *Record of the Expedition to Abyssinia*.

80 J. Waterhouse, *Report on the Cartographic Applications of Photography as used in the Topographical Departments of the Principal States in Central Europe, with Notes on the European and Indian Surveys* (Calcutta, 1870).

81 Pritchard, 'Photography in Connection with the Abyssinian Expedition', p. 603.

82 See, for example, Royal Engineers, 'Native town of Senafé!', NAM Photos 6510–222, pl. 3; Royal Engineers, 'Abyssinian Thieves in the Stocks at Senafé', NAM Photos 6510–222, pl. 5.

83 See Royal Engineers, 'Antalo Village' and 'Village on Hill, near Ashangi', *Abyssinian Expedition*, FCOL Ethiopia/1, pls. 23, 31.

84 Royal Engineers, 'Martello Tower, Near Adabaga (photograph)', *Abyssinian Expedition*, FCOL Ethiopia/1, pl. 20; Mr Holmes, 'Martello Tower, Near Adabaga (sketch)', *Abyssinian Expedition*, FCOL Ethiopia/1, pl. 21.

85 See Bates, *Abyssinian Difficulty*, pp. 136–8, 148–51. See also Wilkins, *Reconnoitring in Abyssinia*, p. 83.

86 Royal Engineers, 'Abyssinian Fiddler', *Abyssinian Expedition*, FCOL Ethiopia/1, pl. 64.

87 Markham, *History of the Abyssinian Expedition*, p. 163.

88 T. Baines, *Troops Ascending a Ravine from Annesley Bay in Abyssinia* (1868), oil on canvas, Gubbins Africana Library, in J. Carruthers and M. Arnold, *The Life and Work of Thomas Baines* (Vlaeberg, 1995), p. 101.

89 Callwell, *Small Wars*, p. 38.

90 Wilson, 'Address to the Geographical Section of the British Association', p. 63.

91 *Illustrated London News*, 20 June 1868, cited in Carruthers and Arnold, *The Life and Work of Thomas Baines*, p. 100.

92 See Porter, *History of the Corps of Royal Engineers*, I, p. 5.

93 R. I. Murchison, 'Presidential Address', *PRGS*, XII (1868), p. 275.

94 See *PRGS*, XII (1868), p. 174.

95 Murchison, 'Presidential Address', p. 275.

96 MacKenzie, ed., in *Popular Imperialism and the Military*, p. 4; Stafford, *Scientist of Empire*, p. 209.

97 Markham, 'Geographical Results of the Abyssinian Expedition', p. 49.

98 See discussion after ibid., pp. 115–19.

99 See T. Pakenham, *The Scramble for Africa 1876–1912* (London, 1992), p. xxvii.

100 W. E. Fry, *Occupation of Mashonaland* (1890). Fifty original albums, each with 150 carbon prints, were produced: see RCS Photos Y3052 A; NAM Photos 8206–103.

101 See, for example, Fry, 'The Administrator and Civil Staff', 'Pioneer Officers' and 'Police Officers', in ibid., RCS Photos Y3052 A (1–23).

102 See, for example, Fry, 'Police Tents, Tuli River', ibid., RCS Photos Y3052 A (40).

103 Markham, 'Geographical Results of the Abyssinian Expedition', p. 49.

104 See, for example, E. Lee, *To the Bitter End: A Photographic History of the Boer War, 1899–1902* (London, 1985).

4 *Hunting with the Camera*

1 S. Sontag, *On Photography* (London, 1979), pp. 14–15.

2 M. Brander, *The Big Game Hunters* (London, 1988), pp. 9–11; B. Bull, *Safari: A Chronicle of Adventure* (London, 1992).

3 L. Barber, *The Heyday of Natural History* (New York, 1980).

4 W. Beinart, 'Empire, Hunting and Ecological Change in Southern and Central Africa', *Past and Present*, CXXVIII (1990), pp. 162–86.

5 H. Ritvo, *The Animal Estate: The English and Other Creatures in the Victorian Age* (Harmondsworth, 1990), pp. 203–88.
6 J. M. MacKenzie, *The Empire of Nature: Hunting, Conservation and British Imperialism* (Manchester, 1988).
7 See, for example, W. K. Storey, 'Big Cats and Imperialism: Lion and Tiger Hunting in Kenya and Northern India, 1898–1930', *Journal of World History*, II (1991), pp. 135–73.
8 W. C. Harris, *The Wild Sports of Southern Africa* (London, 1838). By 1852 Harris's account was in its fifth edition.
9 W. C. Harris, *Portraits of the Game and Wild Animals of Southern Africa* (London, 1840).
10 W. W. Hooper and V. S. G. Western, 'Tiger Shooting' (*c.* 1870), RGS Photos E119/015651–015662.
11 See J. Falconer, 'Willoughby Wallace Hooper: "A Craze About Photography" ', *Photographic Collector*, IV (1984), pp. 258–86.
12 R. Barthes, *Camera Lucida: Reflections on Photography* (London, 1984), p. 15.
13 R. Ormond, *Sir Edwin Landseer* (London, 1981).
14 T. R. Pringle, 'The Privation of History: Landseer, Victoria and the Highland Myth' in *The Iconography of Landscape*, ed. D. Cosgrove and S. Daniels (Cambridge, 1988), pp. 142–61; MacKenzie, *Empire of Nature*, pp. 31–4.
15 See ibid., pp. 167–99.
16 G. N. Curzon, *British Government in India: The Story of the Viceroys and Government Houses* (London, 1925), I, p. 258.
17 See, for example, Diyal & Sons, 'Her Excellency in the Jungle', 'Their Excellencies in the Jungle' and 'Her Excellency Crossing a Nullah', from Lala Din Diyal & Sons, *Souvenir of the Visit of HE Lord Curzon of Kedleston, Viceroy of India to HH the Nizam's Dominions, April 1902*, IOL Photos 430/33 (13, 18 and 28).
18 Diyal & Sons, 'First Tiger Shot by HE Lord Curzon in India, Gwalior', from *HE Lord Curzon's First Tour in India* (1899), IOL Photos 430/17 (21).
19 T. Altick, *Shows of London* (London, 1978), p. 299.
20 See also J. Barrell, *The Infection of Thomas De Quincey* (London, 1991), pp. 48–66.
21 V. G. Kiernan, *The Lords of Human Kind: Black Man, Yellow Man, and White Man in an Age of Empire* (London, 1969), p. 64.
22 Curzon, *British Government in India*, I, pp. 103–4, 125.
23 F. V. Emery, 'Geography and Imperialism: The Role of Sir Bartle Frere (1815–84)', *GJ*, CL (1984), pp. 342–50.
24 S. Alexander, *Photographic Scenery of South Africa* (1880), FCOL Photos, South Africa 1.
25 See H. Ricketts, 'Early Indian Photographs at the Graves Art Gallery, Sheffield', *Creative Camera*, CCVIII (1982), pp. 476–84.
26 Alexander, *Photographic Scenery of South Africa*, preface.
27 Ibid., pl. 99: 'Natal – Near Maritzburg'; pl. 100: 'Scene in a Bush'.
28 F. Galton, *Tropical South Africa* (London, 1853).
29 Ibid., pp. 69–70.
30 C. J. Andersson, *Lake Ngami: Explorations and Discoveries During Four Years' Wanderings in the Wilds of South Western Africa* (London, 1856).
31 F. Galton, *The Art of Travel: Or, Shifts and Contrivances Available in Wild Countries* (London, 1855).
32 H. A. Bryden, *Gun and Camera in Southern Africa: A Year of Wanderings in Bechuanaland, the Kalahari Desert, and the Lake River Country, Ngamiland* (London, 1893), p. 535.
33 W. E. Oswell, *William Cotter Oswell: Hunter & Explorer* (London, 1900), 2 vols.
34 J. Forsyth, *The Highlands of Central India: Notes on Their Forests and Wild Tribes, Natural History and Sport* (London, 1871).
35 See, for example, H. W. Seton-Karr, *Ten Years' Wild Sport in Foreign Lands* (London, 1889); M. W. H. Simpson, 'Shooting in the Barbary States' in *Big Game Shooting in Africa*, ed. H. C. Maydon (London, 1932), pp. 122–6.
36 F. C. Selous, 'Twenty Years in Zambesia', *GJ*, I (1893), pp. 289–324.

37 F. C. Selous, *Travel and Adventure in South-East Africa* (London, 1893), p. ix.
38 F. C. Selous, 'Introduction' in C. H. Stigand and D. D. Lyell, *Central African Game and Its Spoor* (London, 1906), pp. xi–xii.
39 Mashonaland and Matabeleland were areas in what became known, by 1895, as 'Rhodesia'.
40 Selous, *Travel and Adventure*, p. 383.
41 See K. Tidrick, *Empire and the English Character* (London, 1990), pp. 48–87.
42 See, for example, F. C. Selous, 'The History of the Matabele, and the Cause and Effect of the Matabele War', *PRCI*, XXV (1893–4), pp. 251–90.
43 W. E. Fry, *Occupational of Mashonaland* (1890), RCS Photos Y3052 A; NAM Photos 8206–103.
44 Fry, 'Sea Cow's Head (shot in Lundi)', ibid., RCS Photos Y3052 A (105).
45 Selous, 'Twenty Years in Zambesia', pp. 277, 321.
46 Brander, *Big Game Hunters*, pp. 144–51.
47 Anon., 'Obituary: Major Chauncey Hugh Stigand', *GJ*, XXVIII (1920), pp. 237–9.
48 C. H. Stigand, *Scouting and Reconnaissance in Savage Countries* (London, 1907), pp. 74–5, 83, 87–8; 129.
49 C. H. Stigand, *Hunting the Elephant in Africa. And Other Recollections of Thirteen Years' Wanderings* (New York, 1913), pp. 309–24.
50 Ibid., p. 311.
51 Ibid., p. 310.
52 Stigand and Lyell, *Central African Game*, p. 1.
53 T. Roosevelt, 'Introduction' in Stigand, *Hunting the Elephant*, p. xi.
54 C. H. Stigand, *An African Hunter's Romance*, unpublished typescript (n.d.), 400 pp., RGS Archives AR 64, 2.
55 L. Davidoff and C. Hall, *Family Fortunes: Men and Women of the English Middle Class, 1780–1850* (London, 1987), p. 406.
56 See L. Tickner, *The Spectacle of Women: Imagery of the Suffrage Campaign 1907–14* (London, 1987).
57 F. A. Dickinson, *Big Game Shooting on the Equator* (London, 1908), pp. 222, 245.
58 E. G. Lardner, *Soldiering and Sport in Uganda* (London, 1912), p. 85.
59 M. Cocker, *Richard Meinertzhagen: Soldier, Scientist and Spy* (London, 1989), pp. 72–3, 172–3.
60 D. Birkett, *Spinsters Abroad: Victorian Lady Explorers* (Oxford, 1989).
61 S. Mills, *Discourses of Difference: An Analysis of Women's Travel Writing and Colonialism* (London, 1991).
62 M. H. Kingsley, *Travels in West Africa: Congo Français, Corsico and Cameroons* (London, 1897), p. 268.
63 A. Blunt, *Travel, Gender, and Imperialism: Mary Kingsley and West Africa* (London, 1994), p. 73.
64 R. B. Loder, 'Journal' (1910–11) and 'British East Africa Journal' (1912–13), RGS Archives.
65 A. Herbert, *Two Dianas in Somaliland: The Record of a Shooting Trip* (London, 1908).
66 Ibid., p. 19.
67 Ibid., dedication.
68 Ibid., p. 205.
69 Ibid., p. 289.
70 Ibid., p. 39.
71 Ibid., p. 63.
72 Ibid., p. 97.
73 C. H. Stigand, *To Abyssinia Through an Unknown Land: An Account of a Journey Through Unexplored Regions of British East Africa by Lake Rudolf to the Kingdom of Menelek* (London, 1910), pp. 83–4.
74 See, for example, Herbert, 'The Oryx at Home', *Two Dianas in Somaliland*, facing p. 48.
75 Ibid., pp. 269–70.

76 Ibid., p. 208.
77 Harris, *Wild Sports*. The 'cameleopard' was an early term for the giraffe.
78 J. G. Millais, *A Breath from the Veld* (London, 1899); J. G. Millais, *Wanderings & Memories* (London, 1919).
79 D. Livingstone to Lord Malmesbury, 17 December 1858, in *The Zambesi Expedition of David Livingstone 1858–1863*, ed. J. P. R. Wallis (London, 1956), II, p. 299.
80 J. Chapman, *Travels in the Interior of South Africa 1849–1863: Hunting and Trading Journeys from Natal to Walvis Bay & Visits to Lake Ngami & Victoria Falls*, ed. E. C. Tabler, (Cape Town, 1971).
81 See Chapman, 'The Hunters at Breakfast on Elephant Foot or Trunk', ibid., pl. 5, pp. 10–11.
82 Ibid., II, p. 211.
83 G. C. Dawnay, album of sepia prints, RGS photos C82/006132–301. See Anon., 'Hon. Guy. C. Dawnay: Obituary', *PRGS*, XI (1889), p. 422.
84 Barthes, *Camera Lucida*, p. 79.
85 R. Ward, *The Sportsman's Handbook to Practical Collecting, Preserving and Artistic Setting-up of Trophies and Specimens* (London, 1882, 2nd edn), p. 11.
86 R. Ward, *The Sportsman's Handbook to Collecting, Preserving and Setting-up Trophies and Specimens* (London, 1911, 10th edn), p. 22.
87 Ibid., p. 139.
88 Ibid., p. 138.
89 C. Darwin, *The Expression of the Emotions in Man and Animals* (London, 1872).
90 R. G. G. Cumming, *Five Years of a Hunter's Life in the Far Interior of Southern Africa* (London, 1850), 2 vols.
91 Altick, *Shows of London*, p. 477.
92 Brander, *Big Game Hunters*, pp. 44–8.
93 See, for example, W. H. Schneider, *An Empire for the Masses: The French Popular Image of Africa, 1870–1902* (Westport, 1982), pp. 125–51; MacKenzie, *Empire of Nature*, p. 31.
94 Altick, *Shows of London*, pp. 290–2.
95 See A. E. Coombes, *Reinventing Africa: Museums, Material Culture and Popular Imagination in Late Victorian and Edwardian England* (London, 1994), pp. 63–83.
96 Selous, *Travel and Adventure*, pp. 90–1.
97 C. G. Schillings, *In Wildest Africa* (London, 1907), 2 vols.
98 Bryden, *Gun and Camera in Southern Africa*, pp. 491–2.
99 C. V. A. Peel, *Somaliland* (London, 1900); C. V. A. Peel, *Wild Sport in the Outer Hebrides* (London, 1901).
100 C. V. A. Peel, *Popular Guide to Mr C. V. A. Peel's Exhibition of Big-game Trophies and Museum of Natural History and Anthropology* (Guildford, 1906), pp. 4–6.
101 J. H. Patterson, *The Man-Eaters of Tsavo* (London, 1907).
102 Dickinson, *Big Game Shooting on the Equator*, p. 295.
103 Ward, *Sportsman's Handbook*, p. 14. See also H. C. Maydon, ed., *Big Game Shooting in Africa* (London, 1932), pl. 1: 'Vital Shots on the Elephant'.
104 H. B. George, *The Oberland and its Glaciers: Explored and Illustrated with Ice-Axe and Camera* (London, 1866), p. 3.
105 Ibid., p. iv.
106 N. Broc, *Les montagnes au siècle des lumières: perception et représentation* (Paris, 1991).
107 See Altick, *Shows of London*, pp. 474–7; P. Hansen, 'Albert Smith, the Alpine Club, and the Invention of Mountaineering in Mid-Victorian Britain', *Journal of British Studies*, XXXIV (1995), pp. 300–24.
108 George, *Oberland*, p. 3.
109 H. B. George, 'Photography' in *Hints to Travellers*, ed. G. Back, R. Collinson and F. Galton (London, 1871), pp. 51–3; H. B. George, 'Photography' in *Hints to Travellers*, ed. F. Galton (London, 1878), pp. 47–53.

110 S. Bourne, 'Narrative of a Photographic Trip to Kashmir (Cashmere) and Adjacent Districts', *BJP*, XIII (1866), p. 524.
111 R. J. Fowler, 'Letter from Paris, April 29 1867', *BJP*, XIV (1867), pp. 212–13.
112 Anon., 'A Field for Work', *BJP*, XIV (1868), pp. 119–20.
113 W. Gilpin, 'On Picturesque Travel', from *Three Essays* (1792) in *Nature and Industrialization*, ed. A. Clayre (Oxford, 1977), p. 27.
114 L. Nochlin, 'The Imaginary Orient', *Art in America* (May 1993), p. 127, cited in S. Graham-Brown, *Images of Women: The Portrayal of Women in Photography of the Middle East, 1860–1950* (London, 1988), p. 8.
115 George, *Oberland*, p. 197.
116 Ibid., p. 242.
117 Ibid., p. 192.
118 Ibid., p. 197.
119 H. J. Mackinder, *The First Ascent of Mount Kenya*, ed. M. K. Barbour (London, 1991), p. 31.
120 Ibid., p. 219.
121 H. J. Mackinder, 'A Journey to the Summit of Mount Kenya, British East Africa', *GJ*, XV (1900), pp. 453–86.
122 B. Sharpe to H. J. Mackinder, 20 January 1900, SGO, *GJ*, MP/F/100.
123 Mackinder, 'A Journey to the Summit of Mount Kenya', p. 476.
124 'Mackinder/Hausburg Photographs', SGO, MP/L/100.
125 Mackinder, *First Ascent*, pp. 184, 214.
126 Ibid., pp. 105, 110, 136, 188, 199, 212, 231, 244.
127 Mackinder, 'Journey to the Summit', p. 476.
128 R. B. Sharpe, H. J. Mackinder, E. Saunders and C. Camburn, 'On the Birds Collected During the Mackinder Expedition to Mount Kenya', *Proceedings of the Zoological Society*, III (1900), pp. 596–609; O. Thomas, 'List of Mammals Obtained by Mr H. J. Mackinder During His Recent Expedition to Mount Kenya, British East Africa', *Proceedings of the Zoological Society*, I (1900), pp. 173–80.
129 See Mackinder, *First Ascent*, p. 105.
130 E. N. Buxton, *Two African Trips: With Notes and Suggestions on Big Game Preservation in Africa* (London, 1902), pp. 133–4.
131 Mackinder, *First Ascent*, p. 215.
132 Mackinder, 'Journey to the Summit', p. 454.
133 Mackinder, *First Ascent*, p. 233.
134 A number of animal species collected by Mackinder also bear his name, such as the eagle owl, *Bubo Mackinderi*.
135 See J. W. Arthur, 'Mount Kenya', *GJ*, LVIII (1921), p. 23.
136 F. Haes, 'Photography in the Zoological Gardens', *Photographic News*, X (1865), pp. 78–9, 89–91.
137 F. York, *Animals in the Gardens of the Zoological Society, London, Photographed from Life* (London, 1873).
138 C. Reid, 'Some Experiments in Animal Photography', *BJP*, XXIX (1882), pp. 216–18.
139 E. J. Muybridge, *The Attitudes of Animals in Motion: A Series of Photographs Illustrating the Consecutive Positions Assumed by Animals in Performing Various Movements: Executed at Palo Alto, California, in 1878 and 1879* (San Francisco, Ca., 1881).
140 T. R. Dallmeyer, *Telephotography, an Elementary Treatise on the Construction and Application of the Telephotographic Lens* (London, 1899).
141 J. Coles, ed., *Hints to Travellers: Scientific and General* (London, 1901), I and II.
142 See G. Didi-Huberman, 'Photography-Scientific and Pseudo-Scientific' in *A History of Photography: Social and Cultural Perspectives*, ed. J. Lemagny and A. Rouillé (Cambridge, 1987), pp. 71–5.
143 S. Giedion, *Mechanization Takes Command: A Contribution to an Anonymous History* (London, 1969), pp. 17–30.

144 See, for example, E. Bennet, *Shots and Snapshots in British East Africa* (London, 1914), pp. 268–9.
145 J. E. Cornwall, *Photographic Advertising in England 1890–1960* (Giessen-Wieseck, 1978).
146 C. J. Cornish, ed., *The Living Animals of the World: A Popular Natural History* (London, n.d).
147 D. English, *Photography for Naturalists* (London, 1901); F. C. Snell, *A Camera in the Fields: A Practical Guide to Nature Photography* (London, 1905); E. J. Bedford, *Nature Photography for Beginners* (London, 1909); S. C. Johnson, *Nature Photography: What to Photograph, Where to Search for Objects, How to Photograph Them* (London, 1912).
148 R. Kearton, *Wild Life at Home: How to Study and Photograph It* (London, 1898).
149 C. Kearton, *Wild Life Across the World* (London, 1913).
150 Buxton, *Two African Trips*, p. v.
151 J. M. MacKenzie, *The Empire of Nature* (Manchester, 1988), pp. 211–16.
152 E. N. Buxton, *Short Stalks: Or Hunting Camps North, South, East and West* (London, 1892), p. 96.
153 Buxton, *Two African Trips*, pp. 54, 90.
154 Ibid., p. 106.
155 Ibid., facing pp. 117, 125.
156 Ibid., pp. 90–93.
157 Ibid., p. 2.
158 H. H. Johnston, 'Introduction' to C. G. Schillings, *With Flashlight and Rifle: A Record of Hunting Adventures and of Studies in Wild Life in Equatorial East Africa* (London, 1906), p. xiii.
159 Ibid., p. xiv.
160 Ibid.
161 Schillings, *With Flashlight and Rifle*.
162 See, for example, G. Shiras, 'Photographing Wild Game with Flashlight and Camera', *National Geographic Magazine*, XVII (1906), pp. 367–423.
163 A. R. Dugmore, *Nature and the Camera* (London, 1903).
164 A. R. Dugmore, *Camera Adventures in the African Wilds: Being an Account of a Four Months' Expedition in British East Africa, for the Purpose of Securing Photographs of the Game from Life* (London, 1910); A. R. Dugmore, *Wild Life and the Camera* (London, 1912); A. R. Dugmore, *The Wonderland of Big Game* (London, 1925).
165 Dugmore, *Camera Adventures*, p. xvi.
166 Dugmore, *Wonderland of Big Game*, pp. 9–10.
167 S. F. Harmer, 'Preface' in M. Maxwell, *Stalking Big Game with a Camera in Equatorial Africa* (London, 1925), p. ix.
168 A. L. Butler, 'The Blue Nile & Its Tributaries' in *Big Game Shooting in Africa*, ed. Maydon, p. 131.
169 C. G. Schillings, *In Wildest Africa* (London, 1907), I, pp. 88–9.
170 Ibid., pp. 99–100.
171 Dugmore, *Camera Adventures*, pp. xvi–xvii.
172 D. D. Lyell, *Memories of an African Hunter* (London, 1923), p. 157.
173 Dugmore, *Camera Adventures*, p. 22.
174 See, for example, Schillings's photograph 'When My Bullet Hit It, the Rhinoceros Threw Its Head Several Times', *With Flashlight and Rifle*, p. 229.
175 See 'The Author and his camera', in Dugmore, *Camera Adventures*, facing p. 204.
176 Maxwell, *Stalking Big Game*, p. xxi.
177 Ibid., p. 13.
178 Ibid., Chapter XII, pls. 1–5.
179 C. Kearton, *Photographing Wild Life Across the World* (London, n.d.), p. 15.
180 See, for example, 'Wounded Lions', Dugmore, *Camera Adventures*, facing p. 82, and text pp. 82–4; 'A Wounded Bull Giraffe at Close Quarters', Schillings, *With Flashlight and Rifle*, p. 321.

181 Kearton, *Photographing Wild Life*, p. 14.
182 Dugmore, *Wonderland of Big Game*, p. 13.
183 Dickinson, *Big Game Shooting*, pp. 14–15.
184 G. Babault, *Chasses et récherchés zoologique en Afrique orientale anglaise* (Paris, 1917).
185 C. E. Akeley, *In Brightest Africa* (London, 1924). For an important discussion of Akeley's work, see D. Haraway, *Primate Visions: Gender, Race, and Nature in the World of Modern Science* (London, 1992), pp. 26–58.
186 Ibid., p. 45.
187 Lyell, *Memories*, p. 126.
188 Stigand and Lyell, *Central African Game*, p. 5.
189 Stigand, *Hunting the Elephant*, p. 17.
190 Buxton, *Two African Trips*, p. 40.
191 C. H. Stigand, *The Land of Zinj* (London, 1913), p. 309.
192 Ibid., p. 320.
193 W. R. Foran, *Kill or be Killed: The Rambling Reminiscences of an Amateur Hunter* (1933), p. 76. For Foran's extensive collection of photographs, see RCS Photos Y30469K/99; RHLO: Mss. Afr.S. 771–775.
194 Ibid., p. 9.
195 Ibid., p. 91.
196 Ibid., p. 109.
197 See MacKenzie, *Empire of Nature*, pp. 295–311.

5 *'Photographing the Natives'*

1 Anon., 'The Exhibition of the Photographic Society', *Art Journal*, n.s. VI (1854), p. 49.
2 T. Asad, *Anthropology and the Colonial Encounter* (London, 1973); G. W. Stocking, *Victorian Anthropology* (New York, 1987).
3 Stocking, *Victorian Anthropology*, pp. 78–109.
4 For a more detailed account of this see V. Rae-Ellis, 'The Representation of Trucanini' in *Anthropology and Photography 1860–1920*, ed. E. Edwards (London, 1992), pp. 230–33.
5 See R. Poignant, 'Surveying the Field of View: The Making of the RAI Photographic Collection', ibid., p. 46.
6 See J. Comaroff and J. Comaroff, 'Through the Looking-Glass: Colonial Encounters of the First Kind', *Journal of Historical Sociology*, I (1988), pp. 17–23.
7 J. Thomson, 'Comments on Photography', *PRGS*, IV (1882), p. 212.
8 J. Thomson, 'Note on the African Tribes of the British Empire', *JAI*, XVI (1886), pp. 182–6.
9 J. Thomson, *Through Masai Land* (London, 1887), pp. 46–8.
10 Comaroff and Comaroff, 'Through the Looking-Glass', p. 23.
11 J. Thomson, *Illustrations of China and Its People, a Series of Two Hundred Photographs with Letterpress Description of the Places and People Represented* (London, 1873–4), introduction.
12 M. Taussig, *Mimesis and Alterity: A Particular History of the Senses* (London, 1993), p. 198.
13 J. Thomson, 'Photography' in *Hints to Travellers*, ed. E. A. Reeves (London, 1921), II, p. 53.
14 T. Mitchell, 'The World as Exhibition', *Comparative Studies of Society and History*, XXXI (1989), p. 230.
15 F. Galton, *Narrative of an Explorer in Tropical South Africa: Being an Account of a Visit to Damaraland in 1851* (London, 1889), pp. 53–4. This was originally published as *Tropical South Africa* (1853).
16 M. L. Pratt, *Imperial Eyes: Travel Writing and Transculturation* (London, 1992).
17 S. L. Gilman, *Difference and Pathology: Stereotypes of Sexuality, Race and Madness* (London, 1985), p. 45.
18 See, for example, S. Graham-Brown, *Images of Women: The Portrayal of Women in Photography of the Middle East, 1860–1950* (London, 1988), pp. 41, 137.
19 *Picturing Paradise: Colonial Photography of Samoa, 1875 to 1925*, exhibition catalogue

edited by Casey Blanton: Southeast Museum of Photography, Daytona Beach, Florida; Rautenstrauch-Joest-Museum of Ethnology, Cologne; Pitt Rivers Museum, Oxford; Metropolitan Museum of Art in New York (Florida, 1995).

20 Gilman, *Difference and Pathology*; S. L. Gilman, *Health and Illness: Images of Difference* (London, 1995).

21 Gilman, *Difference and Pathology*, p. 107.

22 M. Banta and C. Hinsley, *From Site to Sight: Anthropology, Photography and the Power of Imagery* (Cambridge, Mass., 1986); Edwards, ed., *Anthropology and Photography 1860–1920*.

23 D. Livingstone to C. Livingstone, 10 May 1858, in *The Zambesi Expedition of David Livingstone*, ed. J. P. R. Wallis (London, 1956), p. 431.

24 R. McKenzie, ' "The Laboratory of Mankind": John McCosh and the Beginnings of Photography in British India', *History of Photography*, XI (1987), pp. 109–18.

25 See S. J. Gould, *The Mismeasure of Man* (Harmondsworth, 1981).

26 E. Edwards, 'Photographic "Types": The Pursuit of Method', *Visual Anthropology*, III (1990), pp. 235–58.

27 D. N. Livingstone, *The Geographical Tradition* (Oxford, 1992), pp. 216–59.

28 D. N. Livingstone, 'Human Acclimatization: Perspectives on a Contested Field of Inquiry in Science, Medicine and Geography', *History of Science*, XXV (1987), pp. 359–94.

29 G. Freund, *Photography and Society* (London, 1980), pp. 55–68.

30 B. V. Street, *The Savage in Literature: Representations of 'Primitive' Society in English Fiction 1858–1920* (London, 1975); M. Cowling, *The Artist as Anthropologist: The Representation of Type and Character in Victorian Art* (Cambridge, 1989).

31 Anon., 'The Exhibition of the Photographic Society', *Art Journal*, n.s. VI (1854), p. 49.

32 E. S. Dallas, 'On Physiognomy', *Cornhill Magazine*, (1861) p. 475, cited in Cowling, *Artist as Anthropologist*, p. 32.

33 F. Galton, *The Art of Travel: Or, Shifts and Contrivances Available in Wild Countries* (London, 1872), p. 2.

34 BAAS, *A Manual of Ethnological Enquiry, Being a Series of Questions Concerning the Human Race* (London, 1852), p. 195. See also J. Urry, '*Notes and Queries on Anthropology* and the Development of Field Methods in British Anthropology, 1870–1920', *Proceedings of the Royal Anthropological Institute of Great Britain and Ireland* (1972), pp. 45–6.

35 BAAS, *Notes and Queries on Anthropology, for the Use of Travellers and Residents in Uncivilized Lands* (London, 1874).

36 E. B. Tylor, 'Anthropology' in *Hints to Travellers*, ed. H. H. Godwin-Austen, J. K. Laughton and D. W. Freshfield (London, 1883), p. 222.

37 E. B. Tylor, 'Presidential Address', *BAAS Report*, LIV (1884), pp. 899–910.

38 See A. R. Hinks, ed., *Hints to Travellers* (London, 1938).

39 See F. Spencer, 'Some Notes on the Attempt to Apply Photography to Anthropometry During the Second Half of the Nineteenth Century' in *Anthropology and Photography 1860–1920*, ed. Edwards, pp. 99–107.

40 J. H. Lamprey, 'On a Method of Measuring the Human Form for the Use of Students in Ethnology', *JESL*, n.s. I (1869), pp. 84–5.

41 Spencer, 'Some Notes on the Attempt to Apply Photography to Anthropometry', pp. 99–107.

42 Lamprey, 'On a Method of Measuring the Human Form', p. 85.

43 C. Pinney, 'The Parallel Histories of Anthropology and Photography' in *Anthropology and Photography 1860–1920*, ed. Edwards, p. 77.

44 Lamprey, 'On a Method of Measuring the Human Form', p. 85.

45 J. H. Lamprey, 'Further remarks on the Ethnology of the Chinese' (unpublished manuscript) (1867), RGS Archives X.437.17.

46 Lamprey, 'On a Method of Measuring the Human Form', p. 85.

47 See RAI Photos 1885–1889, Box 40 (Bonaparte Collection); RAI Photos 5999–6002, Box 159 (N. M. Witt Collection).

48 C. Dammann and F. Dammann, *Ethnological Photographic Gallery of the Various Races of Man* (London, 1876). See also Edwards, 'Photographic "Types" '.
49 T. H. Huxley Papers, Imperial College, London, 1869:30:75, cited in Spencer, 'Some Notes', p. 99.
50 Edwards, 'Photographic "Types" '.
51 Spencer, 'Some Notes on the Attempt to Apply Photography to Anthropometry'.
52 M. V. Portman, 'The Exploration and Survey of the Little Andamans', *PRGS*, X (1888), pp. 567–76.
53 M. V. Portman, *Andamanese Islanders* (1893), IOL Photo 188/1–11.
54 W. H. Flower, 'Address to Department of Anthropology', *BAAS*, LI (1881), pp. 682–9.
55 W. H. Flower, 'On the Osteology and Affinities of the Andaman Islands', *JAI*, IX (1880), p. 132.
56 W. H. Flower, comments after Portman 'The Exploration and Survey of the Little Andamans', p. 576.
57 Portman, *Andamanese Islanders* (1893), IOL photos 188/8 and 9.
58 Ibid., IOL photos 188/9, preface.
59 Ibid., IOL photos 188/1. This contains twenty-five front and profile full-face photographs of men and women.
60 Ibid., IOL photos 188/1(23).
61 Ibid., IOL photos 188/6(17).
62 See E. Edwards, 'Science Visualized: E. H. Man in the Andaman Islands' in *Anthropology and Photography 1860–1920*, ed. Edwards, p. 116.
63 M. V. Portman, 'Photography for Anhthropologists', *JAI*, XXC (1896), p. 77.
64 Portman, *Andamanese Islanders*, IOL photos 188/11 (6–24).
65 See Portman, (1890), RAI 810–834, Box 146.
66 Portman, *Andaman Islanders*, IOL photos 188/6 (13).
67 Portman, 'The Exploration and Survey of the Little Andamans', p. 575.
68 Ibid.
69 See E. H. Man, 'Europeans with a group of Öngés, Little Andaman, 1880s', Edwards, 'Science Visualized', pl. 73.
70 Portman, 'Photography for Anthropologists', p. 77.
71 J. Forbes Watson and J. W. Kaye, *The People of India: A Series of Photographic Illustrations, with Descriptive Letterpress, of the Races and Tribes of Hindustan* (London, 1868–75), I, preface.
72 For Hooper's interest in photography in association with hunting, see pp. 100–101 and illus. 37.
73 Watson and Kaye, *The People of India*, I, preface.
74 J. Lubbock, *The Origin of Civilization and the Primitive Condition of Man* (London, 1870).
75 T. H. Huxley, 'Opening Address', *JESL*, n.s. I (1869), p. 90.
76 *JESL*, n.s. I (1869), p. x.
77 G. Campbell, 'On the Races of India as Traced in Existing Tribes and Castes', *JESL*, n.s. I (1869), pp. 128–42.
78 *Congrès internationale des sciences géographique* (Paris, 1876), II, p. 416.
79 Watson and Kaye, *The People of India*, IV, pl. 192.
80 S. Nigam, 'Disciplining and Policing the "Criminals by Birth" ', *Indian Economic and Social History Review*, XXVII (1990), p. 131.
81 Watson and Kaye, *People of India*, IV, pl. 180.
82 C. Pinney, 'Classification and Fantasy in the Photographic Construction of Caste and Tribe', *Visual Anthropology*, III (1990), pp. 259–88.
83 T. Saunders, 'The First General Census of India', *BAAS Report*, LIV (1884), p. 804.
84 E. W. Said, *Orientalism* (London, 1978), p. 22.
85 R. Oliver, *Sir Harry Johnston and the Scramble for Africa* (London, 1957); J. A. Casada, 'Sir Harry H. Johnston as a Geographer', *GJ*, CXLIII (1977), pp. 393–406.
86 See, for example, H. H. Johnston, *British Central Africa* (London, 1897).

87 H. H. Johnston, *The Uganda Protectorate: An Attempt to Give Some Description of the Physical Geography, Botany, Zoology, Anthropology, Languages and History of the Territories Under British Protection in East Central Africa* (London, 1902).
88 Ibid., I, p. vi.
89 Ibid., p. 2.
90 J. Thomson, 'Notes on Cambodia and Its Races', *JESL* (1867), pp. 246–52.
91 J. Doolittle, *Social Life of the Chinese: A Daguerrotype of Daily Life in China* (London, 1868).
92 Anon., 'Illustrations of China and Its People', *BJP*, XX (1873), p. 570.
93 J. Thomson, 'Male Heads, Chinese and Mongolian', *Illustrations of China and Its People* (London, 1873–4), II, pl. 9.
94 Ibid.
95 Ibid.
96 See *Congrès internationale des sciences géographique*, II, pp. 416, 432.
97 Thomson, 'Male Heads, Chinese and Mongolian'.
98 Thomson, 'Mongols', *Illustration of China*, IV, pl. 17, no. 45.
99 J. Thomson, 'Photography and Exploration', *PRGS*, n.s. XIII (1891), p. 672.
100 Thomson, 'Street Groups, Kiu-kiang', *Illustrations of China*, III, pl. 14, no. 25.
101 Thomson, 'Chinese Medical Men' and 'Dealers in Ancient Bronzes &c.', *Illustrations of China*, IV, pl. 11, nos. 26 and 28.
102 W. Allen, 'The Palaver', *Picturesque Views on the River Niger, Sketched during Lander's Last Visit in 1832–33* (London, 1840), pl. 12.
103 For example, see Edwards, 'Science Visualized', pp. 108–21, pl. 67.
104 Cowling, *Artist as Anthropologist*.
105 See Anon., 'Obituary: E. Delmar Morgan', *GJ*, XXIV (1909), pp. 94–5. See also RGS PR/029328–029466.
106 C. Mackerras, *Western Images of China* (Oxford, 1989); L. C. Goodrich and N. Cameron, *The Face of China as Seen by Photographers and Travellers 1860–1912* (Philadelphia, 1978).
107 Thomson, 'Small Feet of Chinese Ladies', *Illustrations of China*, II, pl. 14, no. 39.
108 Thomson, 'Opium Smoking in a Restaurant' and 'A Whiff of Opium at Home', ibid., I, pls. 9 and 10.
109 V. G. Kiernan, *The Lords of Human Kind: Black Man, Yellow Man, and White Man in an Age of Empire* (London, 1969), p. 162.
110 J. Thomson, *The Straits of Malacca, Indo-China and China or Ten Years' Travels, Adventures and Residence Abroad* (London, 1875).
111 Ibid, introduction.
112 Thomson, 'A Pekingese Costermonger', *Illustrations of China*, IV, pl. 11, no. 29. See also Thomson 'Chinese Medical Men', ibid., IV, pl. 11, no. 26.
113 J. Taylor, 'The Alphabetic Universe: Photography and the Picturesque Landscape' in *Reading Landscape: Country-City-Capital*, ed. S. Pugh (Manchester, 1990), p. 184.
114 Stocking, *Victorian Anthropology*, p. 261.
115 BAAS, 'Report of the Anthropometric Committee', *BAAS Report*, L (1880), p. 120. For an important account see Poignant, 'Surveying the Field of View' in *Anthropology and Photography 1860–1920*, ed. Edwards, pp. 58–60.
116 Ibid., p. 121.
117 P. B. Rich, *Race and Empire in British Politics* (Cambridge, 1986), pp. 12–26.
118 BAAS, 'Report of the Anthropometric Committee' (1880), p. 120.
119 BAAS 'Recommendations Adopted by the General Committee', *BAAS Report*, LI (1881), pp. lxiii–lxix. This new committee included William Flower, John Beddoe, F. Brabrook, Francis Galton, J. Park Harrison, Charles Roberts and General Pitt Rivers.
120 BAAS, 'First Report of the Committee', *BAAS Report*, LII (1882), p. 270.
121 Ibid.
122 J. B. Davis and J. Thurnam, *Crania Britannica. Delineations and Descriptions of the Skulls of*

the Aboriginal and Early Inhabitants of the British Isles, 2 vols (London, 1865). See BAAS, 'Report of the Committee', *BAAS Report*, LIII (1883), pp. 306–8.

123 BAAS Racial Committee Albums: I, II and III, (*c.* 1883) RAI Photos. These albums contain 557 portraits from the 884 photographs thought to have been amassed by the committee.

124 BAAS 'Report of the Committee' (1883), pp. 306–8.

125 See 'East Coast of Yorkshire' (*c.* 1882), B-type *carte-de-visite*, selected by J. P. Harrison, BAAS Racial Committee Album II, pls. 73–76, RAI 2926–2929.

126 See A. Briggs with A. Miles, *A Victorian Portrait: Victorian Life and Values as Seen Through the Work of Studio Photographers* (London, 1989), pp. 72–91.

127 J. Beddoe, *The Races of Britain: A Contribution to the Anthropology of Western Europe* (Bristol, 1885). Beddoe became President of the Anthropological Society in 1869. He was on the Council of the BAAS from 1870 and was closely involved in the work of its Racial Committees, including the collection of photographs.

128 Beddoe, *Races of Britain*, p. 255.

129 BAAS, 'Final Report of the Anthropometric Committee', *BAAS Report*, LIII (1883), pp. 253–306.

130 Ibid., p. 271.

131 Ibid., pp. 269–71.

132 Ibid., p. 269.

133 Ibid., p. 300.

134 F. Galton, 'On Stereoscopic Maps, Taken from Models of Mountainous Countries', *JRGS*, XXXV (1865), p. 101. Galton later experimented with using the camera to measure the height, distance and rate of movement of clouds. See Galton Manuscripts 118/4, UCL Library.

135 F. Galton, 'Address to the Department of Anthropology', *BAAS Report*, XLVII (1877), pp. 94–100.

136 F. Galton, 'Composite Portraits', *JAI*, VIII (1878), pp. 132–48.

137 Ibid., p. 135.

138 Galton tried subsequently to reverse the process, to isolate the peculiar features of the individual from the common features of the group. See F. Galton, 'Analytical Photography', *Photographic Journal*, XXV (1900), pp. 135–8.

139 For an account which places Galton's photographic work firmly within his science of eugenics, see D. Green, 'Veins of Resemblance: Photography and Eugenics', *Oxford Art Journal*, VII (1984), pp. 3–16.

140 F. Galton, 'On the Application of Composite Portraiture to Anthropological Purposes', *BAAS*, LI (1881), pp. 690–91.

141 F. Galton, 'Exhibition of Composite Photographs of Skulls by Francis Galton', *JAI*, XV (1885), pp. 390–92. See also W. S. Duncan, 'A New Method of Comparing the Forms of Skulls', *BAAS Report*, LIII (1883), pp. 570–71.

142 E. B. Tylor, 'Presidential Address', *BAAS Report*, LIV (1884), pp. 899–910.

143 For one possible response to Tylor, see A. C. Fletcher, 'Composite Portraits of American Indians', *Science* VII (1886), cited in Green, 'Veins of Resemblance', p. 16.

144 F. Galton, *Hereditary Genius* (London, 1869), p. 325.

145 Ibid., p. x.

146 Ibid., pp. 393–5.

147 Ibid., p. xxv.

148 See R. E. Fancher, 'Francis Galton's African Ethnology and Its Role in the Development of His Psychology', *British Journal for the History of Science*, XVI (1983), pp. 67–79.

149 Galton, *Narrative of an Explorer in Tropical South Africa*, preface.

150 Ibid., p. 75.

151 See H. J. Dyos and M. Wolff, eds, *The Victorian City* (London, 1973), 2 vols; P. Keating, ed., *Into Unknown England, 1866–1913: Selections from the Social Explorers* (Glasgow,

1976); D. E. Nord, 'The Social Explorer as Anthropologist' in *Visions of the Modern City*, ed. W. Sharpe and L. Wallock (Baltimore, 1987), pp. 122–34.

152 J. Thomson and A. Smith, *Street Life in London* (London, 1878), preface. This contains thirty-seven photographs reproduced in Woodburytype. These were initially issued in twelve monthly episodes (three photographs and accompanying commentaries per part), between February 1877 and January 1878, by Woodbury Permanent Photographic Printing Co., London, followed by publication as a whole.

153 Ibid.

154 Stephen White, *John Thomson: Life and Photographs – The Orient, Street Life in London, Through Cyprus with the Camera* (London, 1985), p. 31.

155 J. Thomson and A. Smith, *Street Incidents* (London, 1881). For an alternative view, see I. Gibson-Cowan, 'Thomson's *Street Life* in Context', *Creative Camera*, CCLI (1985), pp. 10–15.

156 Adolphe Smith joined the Social Democratic Federation in 1884. He was also involved, as an official interpreter, in the International Trade Union Congresses between 1886 and 1905. Especially interested in writing on hygiene and medicine, he was Special Commissioner for the *Lancet* for over forty years.

157 While all the photographs were made by Thomson, many of the stories accompanying them are credited to Adolphe Smith ('A. S.'). Two narratives are acknowledged as the work of Thomson ('London Nomades', February 1877, and 'Street Floods in Lambeth', March 1877). Ten uncredited episodes would appear to have been jointly written.

158 Thomson and Smith, *Street Life*, preface.

159 Adolphe Smith was the son-in-law of Blanchard Jerrold, editor of the liberal *Lloyd's Weekly London Newspaper*.

160 B. Jerrold and G. Doré, *London: A Pilgrimage* (London, 1872).

161 G. Pollock, 'Vicarious Excitements: *London: A Pilgrimage* by Gustave Doré and Blanchard Jerrold, 1872', *New Formations*, IV (1988), p. 28.

162 Douglas Jerrold, Blanchard's father, was Henry Mayhew's father-in-law. Mayhew and Douglas Jerrold were founding editors of *Punch* in 1841.

163 H. Mayhew, *London Labour and the London Poor* (London, 1861), 4 vols. This resulted from a series of articles Mayhew wrote in 1849 and 1850 for the *Morning Chronicle* and under his own auspices after leaving the paper. The combined material was published together in 1851–2 and reissued in four volumes in 1861–2. The following discussion refers to the 1861–2 edition. See G. Himmelfarb, *The Idea of Poverty: England in the Early Industrial Age* (London, 1984), pp. 312–70.

164 Thomson and Smith, *Street Life*, preface.

165 See A. McCawley, 'An Image of Society' in *A History of Photography: Social and Cultural Perspectives*, ed. J. Lemagny and A. Rouillé (Cambridge, 1987), pp. 62–70.

166 G. H. Martin and D. Francis, 'The Camera's Eye' in *The Victorian City*, ed. Dyos and Wolff, I, pp. 227–46; M. Wolff and C. Fox, 'Pictures from the Magazines', ibid., II, pp. 559–82.

167 S. Spencer, *O. G. Rejlander: Photography as Art* (Ann Arbor, 1985).

168 J. M. Da Costa Nunes, 'O. G. Rejlander's Photographs of Ragged Children', *Nineteenth Century Studies*, IV (1990), pp. 105–36.

169 Da Costa Nunes, 'O. G. Rejlander's Photographs', pp. 118–19.

170 Mayhew, *London Labour*, I, preface.

171 Ibid., I, p. 1.

172 Thomson and Smith, *Street Life*, 'London Nomades'.

173 See Thomson, 'Female Coiffure', *Illustrations of China*, IV, pl. 8, no. 19. See also Thomson, 'Mongols', ibid., IV, pl. 17, no. 45, and 'Male Heads, Chinese and Mongolian', ibid., II, pl. 9.

174 Mayhew, *London Labour*, I, p. 2.

175 Ibid., I, p. 43.

176 H. Mayhew and J. Binney, *The Criminal Prisons of London and Scenes of Prison Life* (London, 1862), p. 4.
177 Ibid.
178 Ibid., p. 5.
179 J. Greenwood, *The Wilds of London* (London, 1874). This included twelve detailed drawings by Alfred Concanen.
180 J. Greenwood, *Prince Dick of Dahomey* (London, 1890); *Old People in Odd Places, or The Great Residuum* (London, n.d.); *The Wild Man at Home* (London, n.d.).
181 Greenwood, *Wilds of London*, p. 5.
182 Ibid., p. 7.
183 A. McClintock, *Imperial Leather: Race, Gender and Sexuality in the Colonial Contest* (London, 1995), pp. 132–80, 207–31.
184 BAAS, 'Final Report of the Anthropometric Committee', pp. 272–3.
185 A. Levy, *Other Women: The Writing of Class, Race, and Gender, 1832–1898* (Princeton, 1991).
186 Greenwood, *Wilds of London*, preface.
187 Mayhew, *London Labour*, IV, advertisement.
188 See E. C. Grenville Murray, *Spendthrifts: And Other Social Photographs* (London, 1887); Shadow [pseud.], *Midnight Scenes and Social Photographs: Being Sketches of Life in the Streets, Wynds, and Dens of the City* (Glasgow, 1857); P. Colins, ed., *Dickens: The Critical Heritage* (London, 1970).
189 See, for example, ' "Hookey Alf" of Whitechapel', Thomson and Smith, *Street Life*, November 1877.
190 Jerrold and Doré, *London*; J. R. Walkowitz, *City of Dreadful Delight: Narratives of Sexual Danger in Late-Victorian London* (London, 1992), pp. 15–39.
191 Mayhew and Binney, *Criminal Prisons*, p. 4, emphasis in original.
192 Ibid., p. 7.
193 Jerrold and Doré, *London*, pp. 147–8; P. J. Keating, 'Fact and Fiction in the East End' in *The Victorian City*, ed. Dyos and Wolff, II, pp. 585–602.
194 J. Barrell, *The Infection of Thomas De Quincey: A Psychopathology of Imperialism* (London, 1991), pp. 48–66, 197.
195 For further discussion of the relationships between vision and power, See M. Foucault, *Discipline and Punish: The Birth of the Prison* (Harmondsworth, 1987), p. 187; J. Fabian, *Time and the Other: How Anthropology Makes Its Object* (New York, 1983), pp. 105–41.
196 B. McGrane, *Beyond Anthropology: Society and the Other* (New York, 1989).
197 E. F. im Thurn, 'Anthropological Uses of the Camera', *JAI*, XXII (1893), p. 186.
198 See, for example, P. Emerson and W. C. Moore, *Geography Through the Stereoscope: Teacher's Manual and Student's Field Guide* (London, 1907).
199 See also *Picturing Paradise: Colonial Photography of Samoa, 1875 to 1925*.

6 *Visual Instruction*

1 *The Queen, Her Empire and the English-Speaking World* (London, 1897), p. viii; *The English-Speaking World: Photographic Reproductions of Its Scenery, Cities and Industry* (London, 1896).
2 *Broader Britain: Photographs Depicting the Scenery, the Cities and the Industries of the Colonies and Dependencies of the Crown* (London, 1895), introduction.
3 See, for example, *The English-Speaking World*, pp. 59, 107, 120, 153, 186.
4 H. O. Arnold-Forster, ed., *The Queen's Empire: A Pictorial and Descriptive Record* (London, 1897).
5 Ibid., p. x.
6 *Pictures of Our Empire* (London, 1907).
7 J. M. MacKenzie, *Propaganda and Empire* (Manchester, 1984), pp. 162–6.
8 CO 885/8 Misc. No. 150 (1902), p. 1.

9 See, for example, C. Lucas, *A Historical Geography of the British Colonies* (London, 1887–1925).
10 G. Kearns, 'Halford John Mackinder 1861–1947', *Geographers Biobibliographical Studies*, IX (1985), pp. 71–86; B. W. Blouet, *Halford Mackinder: A Biography* (Texas, 1987).
11 H. J. Mackinder, 'The Teaching of Geography from an Imperial Point of View, and the Use which Could and Should be Made of Visual Instruction', *GT*, vi (1911), pp. 79–86.
12 Ibid., p. 80.
13 H. J. Mackinder, 'On Thinking Imperially', in M. E. Sadler, ed., *Lectures on Empire* (London, 1907), pp. 32–42.
14 CO 885/8 Misc. No. 157 (1903), pp. 1–9.
15 CO 885/17 Misc. No. 188 (1906), pp. 47–9.
16 CO 885/8 Misc. No. 174 (1905), pp. 1–10.
17 H. J. Mackinder, *Seven Lectures on the United Kingdom for Use in India. Revised for Use in the United Kingdom* (London, 1909), pp. v-vii.
18 CO 885/17 Misc. No. 188 (1906), pp. 78.
19 *Who Was Who 1941–1950*, IV (London, 1950), p. 387.
20 CO 885/17. Misc. No. 188 (1907), p. 112.
21 Twenty-eight albums of Fisher's photographs survive, RCS FC/1.
22 CO 885/17 Misc. No. 188 (1907), pp. 134–6.
23 Ibid., p. 136.
24 Ibid. (1905), p. 2.
25 Ibid. (1907), p. 134.
26 Ibid., p. 135.
27 S. Humphries, *Victorian Britain Through the Magic Lantern* (London, 1989).
28 D. Simpson, 'The Magic Lantern and Imperialism', *Royal Commonwealth Society Library Notes* (1973).
29 MacKenzie, *Propaganda and Empire*, pp. 147–72.
30 CO 885/17 Misc. No. 188 (1908), p. 84.
31 CO 885/8 Misc. No. 152 (1902), p. 7.
32 J. S. Keltie, *Report of the Proceedings of the Society in Reference to the Improvement of Geographical Education* (London, 1886), p. 45.
33 J. S. Keltie, 'Catalogue of Exhibition of Educational Appliances used in Geography Education', ibid., pp. 245–343.
34 J. S. Keltie, 'On Appliances used in Teaching Geography', ibid., p. 197.
35 See Freshfield's comments after Keltie, ibid., p. 203.
36 S. Nicholls, 'The Geography Room and Its Essential Equipment', *GT*, VI (1912), pp. 308–14; E. Young, A. R. Laws and M. Byers, 'Geography Class Rooms', *GT*, VI (1912), pp. 314–23.
37 Keltie, 'On Appliances', pp. 193, 199.
38 See, for example, H. E. Roscoe et al., *Science Lectures for the People* (Manchester, 1866–7).
39 See D. W. Freshfield, 'Valedictory Address', *GT*, VI (1911), pp. 5–9.
40 J. S. Keltie and H. R. Mill, *Report of the Sixth International Geographical Congress* (London, 1895), appendix B, pp. 16–17.
41 D. W. Freshfield, 'Presidential Address', *GT*, II (1903), p. 12.
42 H. R. Mill, *The Record of the Royal Geographical Society 1830–1930* (London, 1930), p. 103.
43 H. J. Mackinder, *India: Eight Lectures Prepared for the Visual Instruction Committee of the Colonial Office* (London, 1910). The textbook, with the same title, was published by George Philip & Son, London. Subsequent citations refer to the latter publication.
44 A. H. Fisher, *Through India and Burmah with Pen and Brush* (London, n.d. [1910]), preface.
45 A. H. Fisher to H. J. Mackinder, 13 November 1907, RCS FC/5/i, p. 67.
46 Mackinder, *India*, p. v.
47 Mackinder, *India*, p. 6.

48 T. Mitchell, 'The World as Exhibition', *Comparative Studies of Society and History*, XXXI (1989), pp. 217–36.
49 P. Greenhalgh, *Ephemeral Vistas: The Expositions Universelles, Great Exhibitions and World's Fairs, 1851–1939* (Manchester, 1988), p. 59; C. A. Breckenridge, 'The Aesthetics and Politics of Colonial Collecting: India at World Fairs', *Comparative Studies in Society and History*, XXXI (1989), pp. 195–216.
50 S. Low, *A Vision of India* (London, 1906), preface.
51 C. K. Cooke, 'The Royal Colonial Tour', *Empire Review*, II (1901), p. 556.
52 CO 885/17 Misc. No. 188 (1907), p. 134.
53 See, for example, W. E. Hooper, ed., *The British Empire in the First Year of the Twentieth Century and the Last of the Victorian Reign: Its Capital Cities and Notable Men* (London, 1904), 2 vols.
54 Fisher to Mackinder, no date, RCS FC/5/i, p. 650.
55 Mackinder, *India*, pp. 15–16.
56 For example, Henry Haward, Head Assistant to the Photographic Department, Survey of India (1883–1905), made a similar series of photographs from the same place. See H. Haward, IOL photo 527 (21–25).
57 Mackinder, *India*, pp. 42–3.
58 Ibid., pp. 71–2, 107.
59 For examples of the varied impact of the rebellion on the representation of India, Indians and the British in India, in print, painting and photography, see *The Raj: India and the British 1600–1947*, exhibition catalogue edited by C. A. Bayly: National Portrait Gallery, London (London, 1990), pp. 231–49.
60 I have followed Fisher's use of place-names: for example, Cawnpore rather than today's Kanpur.
61 Arnold-Forster, *The Queen's Empire*, II, p. 263.
62 Mackinder, *India*, p. 63.
63 Ibid., p. 67.
64 See, for example, Fisher to Mackinder, 31 December 1907, RCS FC/5/i, p. 281; Fisher to Mackinder, 10 January 1908, RCS FC/5/i, pp. 387–92.
65 Mackinder, *India*, p. 59.
66 Fisher, notes on album page, RCS FC/1, album 4. Fisher's photograph was also reproduced in Mackinder, *India*, facing p. 54.
67 See, for example, Shepherd and Robertson, 'Udasees (Fakirs), Delhi' (*c*. 1862), RCS Y33022b/96 (1127); 'Snake Charmers' (*c*. 1862), RGS Photos G13/21 (1123).
68 Fisher to Mackinder, 10 January 1908, RCS FC/5/i, p. 336. See also RCS FC/1, album 2, no. 331.
69 Fisher to Mackinder, 10 January 1908, RCS FC/5/i, p. 346.
70 Ibid., pp. 360–61.
71 CO 885/17 Misc. No. 188 (1907), p. 135.
72 See, for example, Fisher to Mackinder, no date, RCS FC/5/i, p. 536.
73 Fisher to Mackinder, 10 January 1908, RCS FC/f/i, p. 374. See also RCS FC/1, album 3, nos 382–5.
74 Fisher to Mackinder, no date, RCS FC/5/i, p. 544.
75 ibid.
76 Mackinder, *India*, pp. 60–61. For Fisher's descriptions, see Fisher to Mackinder, no date, RCS FC/5/i, pp. 549–54.
77 Mackinder, *India*, pp. 60–61.
78 Fisher to Mackinder, no date, RCS FC/5/i, pp. 567–9.
79 Ibid., pp. 567–70.
80 Mackinder, *India*, pp. 68–71.
81 Fisher to Mackinder, no date, RCS FC/5/i, pp. 845–9; RCS FC/1, album 6, nos 875–7.
82 See, for example, Arnold-Forster, *The Queen's Empire*, II, p. 192.
83 Mackinder, *India*, p. 130.

84 CO 885/17 Misc. No. 188 (1907), p. 138.
85 Mackinder, *India*, p. 118.
86 Ibid., p. 13.
87 Ibid., pp. 42–3.
88 Ibid., p. 79.
89 See RCS FC/1, album 6, nos 846–7.
90 Mackinder, *India*, p. 118.
91 Ibid.
92 W. G. Baker, *The British Empire: The Colonies and Dependencies* (London, 1890), p. 49.
93 Fisher, *Through India and Burmah*, p. 347.
94 J. B. R[eynolds], 'Review of H. J. Mackinder, *Seven Lectures on the United Kingdom for Use in India* (1909)' *GT*, VI (1911), p. 73.
95 CO 885/21. Misc. No. 249, p. 41. See also A. J. Sargent, *The Sea Route to the East: Gibraltar to Wei-hai-wei: Six Lectures Prepared for the Visual Instruction Committee of the Colonial Office* (London, 1912); A. J. Sargent, *Canada and Newfoundland: Seven Lectures Prepared for the Visual Instruction Committee of the Colonial Office* (London, 1913); A. J. Sargent, *Australasia: Eight Lectures Prepared for the Visual Instruction Committee of the Colonial Office* (London, 1913).
96 Sargent, *Sea Route to the East*, pp. 43–5.
97 Sargent, *Australasia*, p. 14.
98 Ibid.
99 Sargent, *Canada and Newfoundland*, p. 16.
100 Ibid., p. 85.
101 S. Schama, *Landscape and Memory* (London, 1995), pp. 191–4.
102 Sargent, *Canada and Newfoundland*, p. 101.
103 Ibid., p. 73.
104 Sargent, *Sea Route to the East*, p. 82.
105 Sargent, *Canada and Newfoundland*, p. 90.
106 Sargent, *Sea Route to the East*, p. 123.
107 Sargent, *Australasia*, p. 55.
108 By September 1911, only six complete sets of slides and 914 textbooks of Mackinder's India lectures had been sold: CO 885/21 Misc. No. 265 (1911), p. 5. Although at £50 the whole set of 480 slides was expensive, slides for each lecture could be hired from Newton & Co. for 10/6 per night.
109 Everard im Thurn was an important figure in the practice of anthropological photography in the 1890s and his later role in COVIC from 1911 deserves fuller attention than can be given here. See D. Tayler, ' "Very loveable human beings": The Photography of Everard im Thurn' in *Anthropology and Photography 1860–1920*, ed. E. Edwards (Cambridge, 1992), pp. 187–92.
110 See CO 885/22 Misc. No. 276, p. 86.
111 A. J. Sargent, *South Africa: Seven Lectures Prepared for the Visual Instruction Committee of the Colonial Office* (London, 1914); Sir Algernon Aspinal, *The West Indies: Seven Lectures Prepared for the Visual Instruction Committee of the Colonial Office* (London, 1914). Lectures on the African Colonies, to be written by A. Wyatt Tilby, never materialized.
112 For a theoretical deconstruction of Mackinder's privileging of the visual in his geography, see G. Ó. Tuathail, *Critical Geopolitics: The Politics of Writing Global Space* (London, 1996), pp. 75–110.
113 See T. Jeal, *Baden-Powell* (London, 1989), pp. 363–423.
114 Mackinder, 'The Teaching of Geography', p. 80.
115 Ibid., emphasis in original.
116 Ibid., p. 83.
117 H. J. Mackinder, *Our Own Islands* (London, 1906), pp. 2, vi–vii.

118 H. J. Mackinder, *Lands Beyond the Channel* (London, 1908); H. J. Mackinder, *Distant Lands* (London, 1910); H. J. Mackinder, *The Nations of the Modern World* (London, 1912).
119 Mackinder, 'The Teaching of Geography', p. 80.
120 R. Brown, *The Countries of the World: Being a Popular Description of the Various Continets, Islands, Rivers, Seas and Peoples of the Globe* (London, 1876–81), p. 1.
121 See, for example, W. G. Baker, *Realistic Elementary Geography, Taught by Picture and Plan* (London, 1888); W. G. Baker, *The British Empire: The Home Countries* (London, 1889); W. G. Baker, *The British Empire: The Colonies and Dependencies* (London, 1890).
122 W. Bisiker, *The British Empire* (London, 1909), preface.
123 A. McClintock, *Imperial Leather: Race, Gender and Sexuality in the Colonial Contest* (London, 1995), pp. 207–31.
124 CO 885/21 Misc. No. 249 (1912), p. 4.
125 Fisher to Mackinder, 31 December 1907, RCS FC/5/i, pp. 301–2. See also Fisher, *Through India and Burmah*, p. 23.
126 Fisher to Mackinder, 2 November 1908, RCS FC/5/i, pp. 54–5.
127 CO 885/17 Misc. No. 188 (1907), p. 136.
128 Mackinder, 'The Teaching of Geography', p. 81.
129 Ibid., p. 86.

7 *Towards a Conclusion*

1 W. L. Price, *A Manual of Photographic Manipulation, Treating of the Practice of the Art: and Its Various Applications to Nature* (London, 1858), pp. 1–2.
2 J. Thomson, 'Photography and Exploration', *PRGS*, n.s. XIII (1891), p. 673.
3 E. W. Said, *Orientalism* (London, 1978), p. 20.
4 *The Meteor*, edited by members of Rugby School, XLVII (1913), p. 132.
5 See D. Livingstone, 'Sketch of the Victoria Falls' (1860), RGS Archives. Reproduced in *David Livingstone and the Victorian Encounter with Africa*, exhibition catalogue edited by J. M. Mackenzie: National Portrait Gallery, London (London, 1996), cat. no. 2.20.
6 T. Baines, *The Victoria Falls – Zambesi River: Sketched on the Spot during the Journey of J. Chapman and T. Baines* (London, 1865).
7 G. A. Farini, *Through the Kalahari Desert: A Narrative of a Journey with Gun, Camera, and Note-book to Lake N'gami and Back* (London, 1886), p. ix.
8 G. A. Farini, 'A Recent Journey in the Kalahari', *PRGS*, n.s. VIII (1886), pp. 447–8.
9 F. C. Selous, *Travel and Adventure in South-East Africa* (London, 1893), p. ix.
10 J. Morris, *Farewell the Trumpets: An Imperial Retreat* (Harmondsworth, 1978), p. 349.
11 A. H. Beaven, *Imperial London* (London, 1901), p. 63, cited in S. Daniels, *Fields of Vision: Landscape, Imagery and National Identity in England and the United States* (Oxford, 1993), p. 29.
12 Daniels, *Fields of Vision*, p. 31 and fig. 3.
13 H. O. Arnold-Forster, *The Queen's Empire: A Pictorial and Descriptive Record* (London, 1897); H. O. Arnold-Forster, *Our Great City: Or, London the Heart of the Empire* (London, 1900), pp. 251–60.
14 See D. Ades, *Photomontage* (London, 1981).
15 D. Green, 'On Foucault: Disciplinary Power and Photography', *Camerawork*, XXXII (1985), p. 9; D. Green, ' "Classified Subjects": Photography and Anthropology – the Technology of Power', *Ten:8*, XIV (1984), pp. 30–37; R. McGrath, 'Medical Police' *Ten:8*, XIV (1984), pp. 13–18; John Tagg, *The Burden of Representation: Essays on Photographies and Histories* (London, 1988), p. 118.
16 Forster, *Queen's Empire*, p. x.
17 D. Hebdige, *Hiding in the Light: On Images and Things* (London, 1988), p. 140.
18 J. Crary, *Techniques of the Observer: On Vision and Modernity in the Nineteenth Century* (London, 1990).
19 See, for example, T. Dennett, 'Popular Photography and Labour Albums' in *Family*

Snaps: The Meanings of Domestic Photography, ed. J. Spence and P. Holland (London, 1991), pp. 72–83.

20 B. Porter, *Critics of Empire: British Radical Attitudes to Colonialism in Africa, 1895–1914* (London, 1968).

21 See, for example, H. R. Fox Bourne, *The Aborigines Protection Society: Chapters in Its History*, (London, 1899), pp. 52–60.

22 C. A. Cline, *E. D. Morel 1873–1924: The Strategies of Protest* (Belfast, 1980).

23 See, for example, J. H. Harris, 'Rubber is Death' (London, 1906).

24 See, for example, E. D. Morel, 'Consul Casement's Report on the Condition of the Congo State Territory', *West African Mail* (19 February 1904), pp. 1,182–8.

25 E. D. Morel, *King Leopold's Rule in Africa* (London, 1904), facing p. 48.

26 See J. H. Harris, *Dawn in Darkest Africa* (London, 1912).

27 Anon, *The Bystander*, 8 May 1940.

28 See H. Callaway, *Gender, Culture and Empire* (Oxford, 1988); J. Trollope, *Britannia's Daughters: Women of the British Empire* (London, 1983).

29 See M. Vaughan, *Curing Their Ills: Colonial Power and African Illness* (Cambridge, 1991).

30 N. Thomas, *Entangled Objects: Exchange, Material Culture, and Colonialism in the Pacific* (London, 1991), pp. 83–124; J. M. Gutman, *Through Indian Eyes: Nineteenth and Early Twentieth Century Photography from India* (New York, 1982); J. Comaroff and J. Comaroff, 'Through the Looking-Glass: Colonial Encounters of the First Kind', *Journal of Historical Sociology*, 1 (1988), p. 7.

31 See, for example, J. Fabb, *Victoria's Golden Jubilee* (London, 1987); J. Fabb, *Royal Tours of the British Empire* (London, 1989).

32 See J. Taylor, *A Dream of England: Landscape, Photography and the Tourist's Imagination* (Manchester, 1994).

33 C. A. Lutz and J. L. Collins, *Reading National Geographic* (Chicago, 1993), pp. 39–41.

34 See C. Pinney, 'The Parallel Histories of Anthropology and Photography' in *Anthropology and Photography 1860–1920*, ed. E. Edwards, (London, 1992), pp. 74–95.

35 Morris, *Farewell the Trumpets*, p. 348, note 1.

36 J. Morris, *Fisher's Face* (London, 1995).

37 See J. Hirsch, *Family Photographs: Content, Meaning and Effect* (Oxford, 1981); J. Spence and P. Holland, eds, *Family Snaps: The Meanings of Domestic Photography* (London, 1991). See also Z. Yalland, *Boxwallahs* (London, 1995). For a different deconstruction of a colonial family photograph, see G. Pollock, 'Territories of Desire: Reconsiderations of an African Childhood. Dedicated to a Woman Whose Name was Not Really "Julia" ' in *Travellers' Tales: Narratives of Home and Displacement*, ed. G. Robertson et al. (London, 1994), pp. 63–89.

Bibliography

MANUSCRIPT AND PHOTOGRAPHIC ARCHIVES

The Brenthurst Library, Johannesburg, South Africa

Thomas Baines Papers, MS. 49.
Frederick C. Selous Papers, MS. 57, 58.
James Chapman Papers, MS. 168.

Foreign and Commonwealth Office Library, London

Photographic Views of Blantyre, BCA (1900/01–1905), Photos, Malawi/1.
Royal Engineers, *Abyssinian Expedition* (1868–9), Photos, Ethiopia/1.
J. W. Lindt, *Picturesque New Guinea*, Photos, Papua New Guinea/2.

India Office Library, London

Lala Din Diyal & Sons, *Souvenir of the Visit of HE Lord Curzon of Kedleston, Viceroy of India to HH the Nizam's Dominions, April 1902*, Photo 430/33.
HE Lord Curzon's First Tour in India (1899), Photo 430/17.
H. Haward Collection, Photo 527.
Samuel Bourne, *India, Scenery 1874*, Photo 94/4.
Maurice Vidal Portman, *Andamanese Islanders* (1893), Photo 188/1–11.

National Army Museum, London

Frederick Bremner, *Types of the Indian Army* (*c.* 1880), Photo 5701–26.
Royal Engineers, *Abyssinian Expedition* (1869), Photo 7604–43.
John McCosh, 'Album of 310 photographs' (1848–53), Photo 6204–3.
William Ellerton Fry, *Occupation of Mashonaland* (1890), Photo 8206–103.
Sir Stafford Northcote Collection (*c.* 1870), Photo 6510–222.
J. Ritchie, 'MS Notes on John McCosh' (n.d.), Archives 7910–10.

National Library of Scotland, Edinburgh

John Kirk Photographs (Private Collection): Acc. 9942/40–41.

Rhodes House Library, Oxford

Anti-Slavery Society Papers, Mss. Brit. Emp. s. 24/J46–49.

Royal Anthropological Institute, London

British Association for the Advancement of Science, Racial Committee Albums I, II and III (*c.* 1883).
Bonaparte Collection, Box 40, Photos 1885–1889.
N. M. Witt Collection, Box 159, Photos 5999–6002.
M. V. Portman photographs (1890), Box 146, Photos 810–834.

Royal Commonwealth Society Collections, Cambridge

Fisher Collection (FC)
No. 1: 28 albums of photographs taken or collected by A. H. Fisher.
No. 5: Letters to H. J. Mackinder: i. Outward Journey to India and Burmah, 24 October 1907–March 1908, pp. 1–853.

Royal Geographical Society, London

Photographs
A. F. Beaufort Collection, E117.
Prince Roland Bonaparte, Collection Anthropologique (*c.* 1880), RGS PR/032113–032121.
Guy Dawnay Collection, C80; C82/006132–301.
J. Grant, '27 Photographs of Zanzibar', X73/018784–018810.
F. H. H. Guillemard Collection, E55.
W. W. Hooper and V. S. G. Western, *Tiger Shooting* (*c.* 1870), E119/015651–015662.
Delmar Morgan Collection (*c.* 1875), PR/029328–029466.
Royal Engineers, Abyssinian Expedition, PR/036171–036246.
Miscellaneous Albums: E118, E119, G13.
Miscellaneous Photographs: PR/.

Prints
M. O'Reilly, *Twelve Views in the Black Sea and the Bosphorous*, Day & Son, Lithographers to the Queen, London, 1856. D108/1, 17–30.
James Ferguson (Lithographer), *Views in Abyssinia*, London, Harrison & Sons, 1867. D108/125–136.
Miscellaneous prints: D108/.

Museum
Royal Seal and Locket of King Theodore, 63/120.1; 94/120.1.
Chauncey Hugh Stigand Collection

Manuscripts
Correspondence
Henry Walter Bates; T. D. Forsyth; Francis Galton; James A. Grant; Thomas H. Holdich; Harry H. Johnston; John Scott Keltie; Halford John Mackinder; Clements Markham; Roderick Impey Murchison; Cuthbert Peek; Chauncey Hugh Stigand; John Thomson; James T. Walker; War Office; Charles William Wilson.
Private Papers
R. B. Loder, 'Journal Kept by Reginald B. Loder of Mandwell Hall, Northamptonshire, during his visit to British East Africa, 1910–1911'; 'British East Africa Journal, 1912–1913'.
Chauncey Hugh Stigand, 'Essays on Central Africa by Kusiali', n.d., AR/64, 5 (6).
—— 'An African Hunter's Romance', unpublished typscript, n.d., 400 pp. AR/64, 2.
Unpublished Journal MSS
Jones H. Lamprey, 'Further Remarks on the Ethnology of the Chinese', 1867.
John Thomson, 'Notes of a Journey with H. G. Kennedy through Siam to the Ruins of Cambodia', 1866.
Committee Minutes
Scientific Purposes Committee Minute Book, November 1877–March 1883.

School of Geography, Oxford

Mackinder Papers (MP): Letters, Mount Kenya Expedition, MP/F/100.
Mackinder and Hausburg Photographs, Mount Kenya Expedition, 1899, MPL/100 (91 prints), MP/H/200 (62 Glass Lantern-slides).

University College, London

Francis Galton Papers 118/4; 152/8; 158/2; 190.

Wellcome Institute for the History of Medicine, London

John Thomson Correspondence 1920–22, IC 357.

OFFICIAL RECORDS (PUBLIC RECORD OFFICE)

Colonial Office Papers (CO)

CO 885/8 Miscellaneous No. 150, 1902, Lantern Lectures on the British Empire; Memorandum by M. E. Sadler, 8pp.
CO 885/8 Miscellaneous No. 152, 1902, Lantern Lectures on the British Empire.
CO 885/8 Miscellaneous No. 157, 1903, 'Syllabus of a Course of Seven Lectures, Illustrated by Lantern Slides, on a Journey to England from the East', 9pp. (Superseded by No. 174).
CO 885/9 Miscellaneous No. 172, 1904, 'Lantern Lectures on the United Kingdom for Use in the Colonies; Memorandum', 2pp.
CO 885/9 Miscellaneous No. 174, 1905, 'Syllabus of a Course of Seven Lectures, Illustrated by Lantern Slides, on a Journey to England from the East', 10pp.
CO 885/17 Miscellaneous No. 188, 1908. Correspondence Relating to Visual Instruction (1905–1907), 137pp.
CO 885/18 Miscellaneous No. 200, 1907. Illustrated Lectures on the Colonies, for the Use of Schools in This Country; Memorandum, 2pp.
CO 885/19 Miscellaneous No. 218, 1910. Further Correspondence Relating to Visual Instruction (1908–1909), 56pp.
CO 885/21 Miscellaneous No. 249, 1912. Further Correspondence Relating to Visual Instruction (1910–1911), 93pp.
CO 885/21 Miscellaneous No. 265, 1911. The Visual Instruction Committee. Memorandum by C. P. Lucas, 14pp.
CO 885/22 Miscellaneous No. 276, 1914. Further Correspondence Relating to Visual Instruction (1912–1913), 94pp.
CO 885/23 Miscellaneous No. 303, 1915. Further Correspondence Relating to Visual Instruction (1914–1915), 27pp.

CONTEMPORARY PUBLICATIONS

Periodicals and Newspapers

Art Journal; *Art Union Journal*; *British Journal of Photography*; *Cornhill Magazine*; *Empire Review*; *The Field*; *Geographical Journal*; *Geographical Magazine*; *Geographical Teacher*; *Journal of the African Society*; *Journal of the Anthropological Institute*; *Journal of the Ethnological Society of London*; *Journal of the Royal Geographical Society*; *Macmillan's Magazine*; *Mission Field*; *Photographic Journal*; *Photographic News*; *Proceedings of the Royal Colonial Institute*; *Proceedings of the Royal Geographical Society*; *Proceedings of the Zoological Society*; *Professional Papers of the Corps of Royal Engineers*; *Reports of the British Association for the Advancement of Science*; *The West African Mail.*

Books and Articles

William de W. Abney, *Instruction in Photography: For Use at the SME Chatham*, Chatham, 1871.
William Allen, *Picturesque Views on the River Niger, Sketched during Lander's Last Visit in 1832–33*, London, 1840.
Charles John Andersson, *Lake Ngami: Explorations and Discoveries During Four Years' Wanderings in the Wilds of South Western Africa*, London, 1856.
Anon., 'The Abyssinian Expedition', *British Journal of Photography*, XIV, 1867, p. 389.

—— 'The Application of Photography to Military Purposes', *Nature*, II, 1870, pp. 236–7.
—— 'The Application of the Talbotype', *Art Union, Monthly Journal of the Fine Arts and the Arts, Decorative, Ornamental*, VIII, 1846, p. 195.
—— 'Photography Applied to the Purposes of War', *Art Journal*, VI, 1854, p. 152.
H. O. Arnold-Forster, ed., *The Queen's Empire: A Pictorial and Descriptive Record*, London, Paris and Melbourne, 1897.
Robert S. S. Baden-Powell, *Scouting for Boys: A Handbook for Instruction in Good Citizenship*, London, 1908.
John Beddoe, *The Races of Britain: A Contribution to the Anthropology of Western Europe*, Bristol, 1885 (republished London, 1971).
E. Bennet, *Shots and Snapshots in British East Africa*, London, 1914.
Samuel Bourne, 'Narrative of a Photographic Trip to Kashmir (Cashmere) and Adjacent Districts', *British Journal of Photography*, XIII–XIV, 1866–7; XIII: pp. 474–5, 498–9, 524–5, 559–60, 583–4, 617–19; XIV: pp. 4–5, 38–9, 63–4.
—— 'A Photographic Journey Through the Higher Himalayas', *British Journal of Photography*, XVI–XVII, 1870–71; XVI: pp. 570, 579–80, 603, 613–14, 628–9; XVII: pp. 15–16, 39–40, 75–6, 98–9, 125–6, 149–50.
—— 'Photography in the East', *British Journal of Photography*, X, 1863, pp. 268–70, 345–7.
—— 'Ten Weeks with the Camera in the Himalayas', *British Journal of Photography*, XI, 1864, pp. 50–51, 69–70.
British Association for the Advancement of Science, *Notes and Queries on Anthropology, for the Use of Travellers and Residents in Uncivilized Lands*, London, 1874.
—— 'Report of the Committee . . . Appointed for the Purpose of Carrying out the Recommendations of the Anthropometric Committee of 1880', *Report of the British Association for the Advancement of Science*, LII, 1882, pp. 278–80.
—— 'First Report of the Committee . . . Appointed for the Purpose of Obtaining Photographs of the Typical Races in the British Isles', *Report of the British Association for the Advancement of Science*, LII, 1882, pp. 270–74.
—— 'Report of the Commitee . . . Appointed for the Purpose of Defining the Facial Characteristics of the Races and Principal Crosses in the British Isles, and Obtaining Illustrative Photographs' *Report of the British Association for the Advancement of Science*, LIII, 1883, pp. 306–8.
—— 'Final Report of the Anthropometric Committee', *Report of the British Association for the Advancement of Science*, LIII, 1883, pp. 253–306.
Robert Brown, *The Countries of the World: Being a Popular Description of the Various Continents, Islands, Rivers, Seas, and Peoples of the Globe*, London, 1876–81.
H. Anderson Bryden, *Gun and Camera in Southern Africa: A Year of Wanderings in Bechuanaland, the Kalahari Desert, and the Lake River Country, Ngamiland*, London, 1893.
Richard F. Burton, *The Lake Regions of Central Africa: A Picture of Exploration*, London, 1860, 2 vols.
Edward North Buxton, *Short Stalks: Or Hunting Camps North, South, East and West*, London, 1892.
—— *Two African Trips: With Notes and Suggestions on Big Game Preservation in Africa*, London, 1902.
Charles E. Callwell, *Small Wars: Their Principles and Purpose*, London, 1899.
Abel Chapman, *On Safari: Big-Game Hunting in British East Africa with Studies in Bird-Life*, London, 1908.
James Chapman, *Travels in the Interior of South Africa 1849–1863: Hunting and Trading Journeys, from Natal to Walvis Bay & Visits to Lake Ngami & Victoria Falls*, ed. Edward C. Tabler, Cape Town, 1971.
Melville Clarke, *From Simla Through Ladac and Cashmere, 1861*, Calcutta, 1862.
John Coles, ed., *Hints to Travellers: Scientific and General*, London, 1901.
A. C. Cooke, *Routes in Abyssinia*, London, 1867.

C. J. Cornish, ed., *The Living Animals of the World: A Popular Natural History*, London, n.d.
Roualeyn George Gordon Cumming, *Five Years of a Hunter's Life in the Far Interior of Southern Africa*, London, 1850.
Carl Dammann and Frederick W. Dammann, *Ethnological Photographic Gallery of the Various Races of Man*, London, 1876.
Charles Darwin, *The Expression of the Emotions in Man and Animals*, London, 1872.
John Donnelly, 'On Photography and Its Application to Military Purposes', *British Journal of Photography*, VII, 1860, pp. 178–9.
Justus Doolittle, *Social Life of the Chinese: A Daguerreotype of Daily Life in China*, London, 1868.
Arthur Radclyffe Dugmore, *Camera Adventures in the African Wilds: Being an Account of a Four Months' Expedition in British East Africa, for the Purpose of Securing Photographs of the Game from Life*, London, 1910.
—— *Nature and the Camera*, London, 1903.
—— *Wild Life and the Camera*, London, 1912.
—— *The Wonderland of Big Game*, London, 1925.
Philip Henry Egerton, *Journal of a Tour Through Spiti, to the Frontier of Chinese Thibet, with Photographic Illustrations*, London, 1864.
Philip Emerson and William Charles Moore, *Geography Through the Stereoscope: Teacher's Manual and Student's Field Guide*, London, 1907.
A. Hugh Fisher, *Through India and Burmah with Pen and Brush*, London, n.d. [1910].
William Robert Foran, *Kill: Or Be Killed, the Rambling Reminiscences of an Amateur Hunter*, London, 1933.
Douglas W. Freshfield, 'The Place of Geography in Education', *Proceedings of the Royal Geographical Society*, n.s. VIII, 1886, pp. 698–714.
Francis Galton, *The Art of Travel: Or, Shifts and Contrivances Available in Wild Countries*, London, 1855.
—— 'Composite Portraits', *Journal of the Anthropological Institute*, VIII, 1878, pp. 132–48.
—— *Hereditary Genius*, London, 1869.
—— *Inquiries into Human Faculty and Its Development*, London, 1907.
—— *Narrative of an Explorer in Tropical South Africa: Being an Account of a Visit to Damaraland in 1851*, London, 1889.
—— 'On the Application of Composite Portraiture to Anthropological Purposes', *Report of the British Association for the Advancement of Science*, LI, 1881, pp. 690–91.
—— 'On Stereoscopic Maps, Taken from Models of Mountainous Countries', *Journal of the Royal Geographical Society*, XXXV, 1865, pp. 99–104.
Hereford Brooke George, *The Oberland and its Glaciers: Explored and Illustrated with Ice-Axe and Camera*, London, 1866.
James Greenwood, *The Wild Man at Home or, Pictures of Life in Savage Lands*, London, n.d.
—— *The Wilds of London*, London, 1874.
Frank Haes, 'Photography in the Zoological Gardens', *Photographic News*, X, 1865, pp. 78–9, 89–91.
William Cornwallis Harris, *Portraits of the Game and Wild Animals of Southern Africa*, London, 1840.
—— *The Wild Sports of Southern Africa*, London, 1838.
Agnes Herbert, *Two Dianas in Somaliland: The Record of a Shooting Trip*, London, 1908.
Samuel Highley, 'Hints on the Management of Some Difficult Subjects in the Application of Photography to Science', *Report of the British Association for the Advancement of Science*, XXIV, 1854, pp. 69–70.
—— 'On the Means of Applying Photography to War Purposes in the Army and Navy', *Report of the British Association for the Advancement of Science*, XXIV, 1854, p. 70.
Trevenen J. Holland and Henry M. Hozier, *Record of the Expedition to Abyssinia, Compiled by Order of the Secretary of State for War*, London, 1870.
H. N. Hutchinson, J. W. Gregory and R. Lydekker, eds, *Living Races of Mankind: A Popular Illustrated Account of the Customs, Habits, Pursuits, Feasts and Ceremonies of the Races of*

Mankind throughout the World, London, n.d. [1903].
Everard F. im Thurn, 'Anthropological Uses of the Camera', *Journal of the Anthropological Institute*, XXII, 1893, pp. 184–203.
Harry Hamilton Johnston, *British Central Africa*, London, 1897.
—— *The Negro in the New World*, London, 1910.
—— *The Uganda Protectorate: An Attempt to Give Some Description of the Physical Geography, Botany, Zoology, Anthropology, Languages and History of the Territories Under British Protection in East Central Africa*, London, 1902.
Cherry Kearton, *Photographing Wild Life Across the World*, London, n.d.
—— *Wild Life Across the World*, London, 1913.
Richard Kearton, *Wild Life at Home: How to Study and Photograph It. Fully Illustrated by Photographs Taken Direct from Nature by C. Kearton*, London, 1898.
John Scott Keltie, 'Geographical Education, Report to the Council of the Royal Geographical Society', *Royal Geographical Society, Supplementary Papers*, I, 1886, pp. 443–594.
—— 'On Appliances used in Teaching Geography', in *Report of the Proceedings of the Society in Reference to the Improvement of Geographical Education*, London, 1886, pp. 182–203.
John Kirk, 'The Extent to Which Tropical Africa is Suited for Development by the White Races, or Under Their Superintendence', *Report of the Sixth International Geographical Congress, London 1895*, 1896, p. 526,
Jones H. Lamprey, 'On a Method of Measuring the Human Form for the Use of Students in Ehtnology', *Journal of the Ethnological Society*, n.s. I, 1869, pp. 84–5.
J. Bridges Lee, 'Photography as an Aid to the Exploration of New Countries', *Journal of the African Society*, I, 1901, pp. 302–11.
Nöel-Marie-Paymal Lerebours, *Excursions daguerriennes: représentant les vues et les monuments anciens et modernes les plus remarquables du globe*, Paris, 1841–4.
John William Lindt, *Picturesque New Guinea*, London, 1887.
David and Charles Livingstone, *Narrative of an Expedition to the Zambesi & Its Tributaries and of the Discovery of the Lakes Shirwa & Nyassa, 1858–1864*, London, 1865.
John McCosh, *Advice to Officers in India*, London, 1856.
—— 'On the Various Lines of Overland Communication Between India and China', *Proceedings of the Royal Geographical Society*, 1860, pp. 47–54.
Halford John Mackinder, 'A Journey to the Summit of Mount Kenya, British East Africa', *Geographical Journal*, XV, 1900, pp.- 453–86.
—— *The First Ascent of Mount Kenya*, edited with an introduction by Michael K. Barbour, London, 1991.
—— 'Geography in Education', *Geographical Teacher*, II, 1903, p. 100.
—— *India: Eight Lectures Prepared for the Visual Instruction Committee of the Colonial Office*, London, 1910.
—— 'On Thinking Imperially' in *Lectures on Empire*, ed. M. E. Sadler, London, 1907, pp. 32–42.
—— 'The Teaching of Geography from an Imperial Point of View, and the Use Which Could and Should be made of Visual Instruction', *Geographical Teacher*, VI, 1911, pp. 79–86.
Clements R. Markham, *A History of the Abyssinian Expedition*, London, 1869.
—— 'Geographical Results of the Abyssinian Expedition', *Journal of the Royal Geographical Society*, XXXVIII, 1868, pp. 12–49.
Marius Maxwell, *Stalking Big Game with a Camera in Equatorial Africa, with a Monograph on the African Elephant*, London, 1925.
Henry Mayhew, *London Labour and the London Poor*, London, 1861.
Hugh Robert Mill, *The Record of the Royal Geographical Society 1830–1930*, London, 1930.
J. A. da Cunha Moraes, *African Occidental: Album Photographico e Descriptivo*, Lisboa, 1885–8.
R. Whitworth Porter, *History of the Corps of Royal Engineers*, Chatham, 1889, 2 vols.
Maurice Vidal Portman, 'The Exploration and Survey of the Little Andamans', *Proceedings of the Royal Geographical Society*, X, 1888, pp. 567–76.
—— 'Photography for Anthropologists', *Journal of the Anthropological Institute*, XXV, 1896, pp. 75–87.